Teaching and Research in Mathematics

Teaching and Research in Mathematics
A Guide with Applications to Industry

Parisa Fatheddin

CRC Press
Taylor & Francis Group
Boca Raton London New York

CRC Press is an imprint of the
Taylor & Francis Group, an **informa** business

A CHAPMAN & HALL BOOK

First edition published 2022
by CRC Press
6000 Broken Sound Parkway NW, Suite 300, Boca Raton, FL 33487-2742

and by CRC Press
4 Park Square, Milton Park, Abingdon, Oxon, OX14 4RN

CRC Press is an imprint of Taylor & Francis Group, LLC

Library of Congress Cataloging-in-Publication Data
Names: Fatheddin, Parisa, author.
Title: Teaching and research in mathematics : a guide with applications to industry / Parisa Fatheddin.
Description: First edition. | Boca Raton : Chapman & Hall/CRC Press, 2022. | Includes bibliographical references and index.
Identifiers: LCCN 2021062218 (print) | LCCN 2021062219 (ebook) | ISBN 9781032289106 (hardback) | ISBN 9781032289113 (paperback) | ISBN 9781003299073 (ebook)
Subjects: LCSH: Mathematics–Study and teaching. | Mathematics–Study and teaching–Research.
Classification: LCC QA11.2 .F38 2022 (print) | LCC QA11.2 (ebook) | DDC 510.71/1–dc23/eng20220415
LC record available at https://lccn.loc.gov/2021062218
LC ebook record available at https://lccn.loc.gov/2021062219

ISBN: 978-1-032-28910-6 (hbk)
ISBN: 978-1-032-28911-3 (pbk)
ISBN: 978-1-003-29907-3 (ebk)

DOI: 10.1201/9781003299073

Typeset in Palatino
by SPi Technologies India Pvt Ltd (Straive)

To Graduate students in Mathematics and

others that may find the content helpful.

Contents

Preface ix
Biography xi
Introduction xiii

Part I Teaching and Conducting Research   1

1. Teaching 3

2. Research and the Publication Process 17

3. Conferences and Collaborations 29

4. Transition to Reading and Writing Research Papers 41
   4.1   Common words and Expressions for a Mathematics
         Research Paper 48

5. Preparing Documents in Latex 51
   5.1   The Beginning Code 51
   5.2   Typing the Main Content 55
   5.3   Inserting Pictures 66
   5.4   References 67
   5.5   Dissertations 70
   5.6   Beamer for Presentations 72

Part II Modern Applications in Industry   81

6. Schrödinger Equation in Laser Technology 83
   6.1   Schrödinger Equation Modeling Laser Propagation 87
   6.2   Optimal Control 98
   6.3   Phase Screens 117
   6.4   Appendix 123

7. PageRank in Google 129
   7.1   Spectrum of the Google Matrix 131
   7.2   Computing the PageRank 137
   7.3   Improving the PageRank 147

7.4   Appendix 154
    7.4.1   Stochastic Processes.. 154
    7.4.2   Markov Process and Markov Chain 155
    7.4.3   Properties of Markov Chains... 155
    7.4.4   Types of Markov Chains 156
    7.4.5   Computing the Transition Probabilities.................................... 157
    7.4.6   Stationary Probability Distribution 158

8. Stochastic Navier-Stokes Equations in Fluids.................................... 161
  8.1   Stochastic Navier Stokes Equation... 168
  8.2   Large Deviations 180
  8.3   Large Deviations for Stochastic Navier Stokes Equation 191
  8.4   Appendix 201
    8.4.1   Brownian Motion, Lévy Process and Random Walk.......... 201
    8.4.2   Applications of large deviations.. 202

Index.. 207

Preface

During my recent experience in obtaining a PhD in mathematics and completing postdoctoral fellowships, I have found that the information needed for one to successfully go through the process is not openly available. One has to ask around and look at many different online sites to attain the help required. The purpose of this manuscript is to provide that help and information in one source and offer graduate students advice and techniques.

In addition to explaining the structure of teaching and research in mathematics, this manuscript tries to answer the "how" in the process. How to teach to engage students, how to read research papers effectively, how to write a research article in a step-by-step process and how to attend conferences and collaborate with others. The challenges that one is likely to face in research are noted here, and advice is given on ways to overcome them. A chapter is also devoted to offering a detailed introduction to the computer program Latex that is used by the mathematics research community to type research papers.

Noting the importance of applications in modern research, Part II of the book introduces topics that are frequently applied in modern industries. They include laser technology, PageRank in Google and fluid dynamics. I have chosen recent research papers in these three areas and have explained their results that use the material one typically studies in graduate-level mathematics. All papers explained have been published after year 2000. The aim here is to expose the reader to modern applications and help the transition to reading research papers go more smoothly.

This manuscript is intended to serve as a guide for graduate students completing their graduate-level courses and beginning their research in mathematics and for anyone pursuing a career in mathematics or related fields. It is meant to act as a complementary advisor to one's official PhD advisor and help graduate students in the process.

Biography

Parisa Fatheddin is Senior Lecturer at the Ohio State University, Marion, where she teaches various courses and conducts research in mathematics. Her main area of research is stochastic partial differential equations, in which she has contributed to the literature by publishing results on stochastic Navier-Stokes and Schrödinger equations in well-established journals in stochastic analysis. Dr. Fatheddin is also active in research in applied fields such as optics and cyber security. As a visiting assistant professor at the Air Force Institute of Technology, she became involved in an optics laboratory and made a publication on her research results in the journal of *Optics Communications*. She also held a visiting position at the University of Pittsburgh, where she collaborated with well-known mathematicians in fluid dynamics and taught many courses. In the field of cyber security, Dr. Fatheddin has contributed a chapter to the textbook *Mathematics in Cyber Research*, explaining modern research that uses differential equations to model the spread of computer virus. Dr. Fatheddin has taught for ten years and continues to help undergraduate and graduate students find their purpose in studying mathematics.

Introduction

To compose research articles and become accustomed to their structure require great effort and dedication and are major steps taken in a graduate program. Here our intention is to help this transition from graduate studies and make it less difficult and time-consuming. In Part I we discuss helpful techniques on teaching and conducting research, and in Part II we introduce some modern research in mathematics in various industries.

Our goal in Part I is to provide the complete information a mathematics graduate student needs to begin the research process and to find ways to teach effectively. We have devoted the first chapter to teaching and have provided ideas on how to engage and empower students by helping them participate and learn actively. Chapter 2 concentrates on the structure of mathematics research articles and the overall publication process. Chapter 3 describes the setting and structure of research conferences and offers helpful advice on how to give a talk and collaborate with others in the research community. The challenges one is likely to face in research and ways to overcome them are discussed in Chapter 4. An introduction to the computer program Latex, which is typically used to type and prepare articles and dissertations in mathematics, is given in Chapter 5. This chapter outlines the structure and rules in Latex to get one started on the preparation of his/her first drafts.

Often in a graduate mathematics program, the focus tends to be on pure mathematics and students seldom become exposed to applications in engineering and other fields. Applications help strengthen results in pure mathematics and emphasize their importance. The balance between pure and applied mathematics is essential and helps one find more meaning and purpose for his/her work in research. In Part II we focus on modern research of mathematics in laser technology, Google and fluid dynamics. We explain how the concepts one studies in graduate-level mathematics are applied and help the reader observe the connections. We have chosen some recently published articles in each of the three applied areas to explain and discuss the ideas. Chapter 6 provides an introduction to the study of lasers and concentrates on results on nonlinear Schrödinger-type equations used to model laser beam propagation in random media.

Chapter 7 explains some modern developments in PageRank that is used in Google to determine the order by which to list the websites related to a search inquiry. An introduction to research on a fluid model, Navier-Stokes equation, is given in Chapter 8 and results on the well-posedness and asymptotic behavior of its stochastic counterpart are explained.

The aim in Part II is to expose the reader to modern applications and help him/her become acquainted with research papers and how to read and understand them. It helps if the reader attains these papers and follows along. The background and terminology needed for each topic are provided within the corresponding chapter, and an Appendix is given at the end of each chapter to recall the concepts used. Understanding these papers may serve as a gateway to more theoretical papers in mathematics and help one build a foundation to start on research articles. They offer interesting and important applications that can empower one's research. The applications can also be explained to students while teaching a related concept and thus raise the interest of students, as well.

There are many different ways to effectively teach and conduct research, and to be successful, one has to make his/her own observations. Here we offer our observations and advice to aid the reader to reflect on the process and help develop his/her own methods. We also hope to bring some light to important applications of graduate-level theoretical mathematics and walk the reader through some recent articles. We hope that this book acts as an aid to one who is in the transition process from graduate studies to teaching and conducting research in mathematics.

Part I

Teaching and Conducting Research

1

Teaching

Teaching is an important part of being a professor. One has to gain the trust of the students by his/her knowledge and understanding in the subject and act as a role model to get the students through the challenges faced in the course. Two main values of good teaching are caring and accepting its responsibility. Caring so that one does not just lecture and go through the material but slows down to ensure that all students are following and getting motivated and mentally engaged and stimulated. Taking the responsibility so that one works to make sure that each student gets the most out of the course, becomes prepared for the next level and that every student is treated fairly and gets the right outcome at the end based on his/her efforts during the course. In this chapter, I explain how I teach and give suggestions that may help others. Teaching should not be seen as a distraction to your work, but as a purpose to your work. In knowledge, one needs to have followers and be kind to them.

Having structure and making a routine for the course are crucial to the students' learning. In the first session, I hand out a tentative schedule along with the syllabus. The syllabus serves as a contract between you and the students. It must offer all the information a student needs to start the course and act as a guideline. It contains information such as the time and location of the class, course description, prerequisites and the expectations for the course. The syllabus should also remind the students of the university's honor code and include contact information regarding services for those having disabilities and the accommodations that have to be made according to the university. How homework and exams' grades are distributed is also important and has to be provided and not be changed afterwards during the semester. I usually put the image of the university's logo on the first page to make it more attractable and connectable to the students. These logos can easily be found online and inserted in the document. At the end of this chapter, in Figures 1.1–1.3, I have provided parts of one of my syllabi and tentative schedule that may be observed as examples.

As for the tentative schedule, it is essential to be structured and provide students a routine to get accustomed to. The tentative schedule that I provide for my classes, has every day that the class meets and what will be covered on that day along with the homework problems that are assigned for each section and the due dates for each problem set. It gives the students a structure to follow and plan ahead for the exams. I make the homework due on Mondays

DOI: 10.1201/9781003299073-1

so that the students can work on them during the week and have the chance to complete them during the weekend. Making the due dates be consistent on one day of the week is easier for students to get used to. Unlike the syllabus, this schedule may be tentative and can be changed with an updated version given later in the course, if necessary. Not only is the schedule helpful to the students, it also gives the instructor a plan for each week and by composing it before the start of the semester, one can get a good overview of the course at the beginning and be reminded of the topics that need to be covered.

In the first week or two of the semester, I try to adjust to the atmosphere of the class and mindsets and calm the students' anxiety and feeling of tense or mistrust. After the first quiz or homework, they tend to gradually trust and listen and follow your mindset and material. Keep your mind on the material and deliver with confidence. I usually set my office hours before class time and devote all its time to the class. If no student comes, I only focus on the subject I am going to teach and either look over my lecture notes or do problems for their upcoming homework so to post solutions online after grading them. I prepare lecture notes for the week in the weekend before and look over them in the night before each session and in office hours. It is best to avoid working on your research during the hour before the class.

When preparing lecture notes, I use the textbook that the class is using, and the textbook that I myself used to learn the material and sometimes I also use a book from a lower level and a book from a higher level than their textbook. These give me different perspectives to write my own lecture notes and share with the students various examples that are closely related to what they have in their book. It is important to make their textbook the main source, since referring to other books or bringing in not closely related material would likely to distract and confuse the students. Even though the society tends to evolve much faster than the textbooks, it is important to teach at the same level and language of the textbook used. Distrusting or devaluing the textbook, confuses the students since they are at the stage of learning, absorbing and trusting information. Outside material such as applications, can be used to bring more color to the material; however, the main source of the study should be the textbook and the lecture notes should not deviate much from its content and level. The past accomplishments of the instructor in mathematics inspires the students to learn; however, the students are there to learn the specific information in their books and be taught that particular mindset and level.

During the course, it helps to give students tips on how to learn and organize the concepts and explain ways to strategically think through the problems or proofs. An effective instructor, not only knows the information but can also transmit the mindset and teach the skills needed. If the lecture notes are only in the formal language of a textbook without discussing how to think in the problems, then students likely will not get into them and will be encouraged to only memorize and not think through the material. It is best not to use the same lecture notes for more than two times for a course. The third time you teach it, rewrite and add and organize the notes so to keep improving the material you teach and refresh your mind. I usually try to avoid teaching one course for more than two semesters in a row, to give my mind a variety of courses to think about and keep my interest in the content of the course.

There is nothing wrong with learning a concept again and revisiting material from the past. It is like a song. If one is asked to state the lyrics of a song, he/she will probably not be able to recall the words, but when the background music of the song is played, the lyrics will automatically come to mind and the person can easily remember the rest of the lyrics. You have to review and remind yourself of the mindset of each course, then you will recall the concepts and the knowledge you obtained in the past. Try to review and remember the material before the start of the semester and then during the course, provide an atmosphere in the class in which students feel comfortable with asking questions and participating in the discussions. If they ask you something that you don't recall or can't answer right at the moment, you can think about it and share the explanation and answer in the following session. If you show that you are nervous in front of them, then they too will be nervous and will not be able to fully listen to you in the classroom and will likely avoid asking questions. If you say a topic of the course is difficult, then they will find that topic very difficult. It is important to share with the students the interest one has in every topic of the course, including applications. I find it best to first cover the sections on theory and concepts and then focus on the sections on applications of those concepts so that the students first have a good foundation before getting into word problems.

Try to make the mathematics you explain be the focus of the attention of the students. It is ok to look at lecture notes during lectures, but to engage the students, your mind has to be on and have to think through the concepts as you speak and do problems for the class. It is helpful to write a few problems without work in lecture notes and do them in front of the students. Even if you doubt and think and don't do them quickly, the students will likely become engaged and help and think with you. If you just write lecture notes and copy them on the board, then you are not really helping them. Don't let the fear of making mistakes and being perfectionist keep you from doing more and being creative. Don't bound your creativity and with it, along with some imagination, try to get the students involved.

Mathematics is about rigor and precision, and one has to take it seriously to learn it well; however, it is not required to be dry and it helps to develop creative ways to help students connect with the material and build their own math world. Lectures can be made more engaging by bringing in applications. For example, for word problems on exams, I try to make the setting to what they can connect to and see in regular life. In the past, instead of using a cylinder in a problem, I have used a flower vase in a cylindrical shape, instead of a tank in a mixing problem, I have used an aquarium in which tap water is mixed with ocean water, instead of ladder problem in the usual setting of the Pythagorean theorem, I have used putting up a Christmas tree against the wall and asking if it fits, and instead of an object bouncing on a spring, I have made the problem in the setting of a playground where a toddler can hang on to an end of a spring and bounce up and down. Bringing the problems to regular life can make the concepts more attainable and interesting. Getting into the math world and only focusing on the math is easier for the mind; however, to better prepare the students for their future, applications need to be discussed and connections be made. Also sharing creative ways to remember a concept can help students develop ways to learn. It is essential for students to learn how to think through the problems and not just learn them.

To engage students in regular undergraduate level mathematics courses, where problem solving is emphasized, I usually begin each lecture with an introduction to the lesson and do some examples on the board, thinking through them as I explain the steps. Then I put problems on the board and ask the students to work on them and solve the problems by themselves. They can look at their notes and talk to their neighbors and ask questions as I walk around the classroom. Afterwards, I randomly call on students to give me steps to the solution of the problems to write on the board and pause and correct them if necessary. This way other students listen and check to see if the answer is correct. In large classrooms, I pick names at random from the name roster. This makes everyone to do the exercises, be engaged in the classroom and their homework will seem less difficult to them.

Hearing other students do problems, instead of just seeing the instructor do them perfectly in front of them, is very important for the students' learning. Passive learning is not effective and the students need to talk about the concepts and steps, pronounce the terms correctly and learn the language of mathematics along with solving problems. Solving problems in class and talking about them, give students the confidence they need for the course, and even if they make a mistake when working on a problem, pointing it out and correcting it, would help the class, since others are likely to make the same mistake and can thus become aware of it. I try to provide an atmosphere in which students feel comfortable with doing exercises in front of the class and try to correct their answers and help them as they say it so not to put them on the spotlight that might cause discomfort.

When teaching an upperlevel undergraduate course, I try to bring in proofs and get them exposed to the mentality of proving results. Depending on the level of the course, I go over the proofs at the right level of difficulty of their textbook, writing "why?" instead of "Proof" before proving it, since even the word proof tends to turn away many students who have not taken many theoretical courses in mathematics. The switch of mentality in mathematics from doing problems by using a specific technique to proving why the technique works can be challenging. After getting used to the mindset of equations and learning a concept and doing exercises applying that particular concept to different problems, students tend to resist the mentality of proofs. It becomes difficult for them to see the purpose of proving a result that seems intuitive based on their past knowledge. Proofs require one to think sequentially and typically involve more than one or two steps. To ease the transition, I give students a list of proofs they need to know before an exam and ask two or three of them on the exam. As for the proofs that are involved and have many steps, I try to explain them on the board and assure the class that they will not be tested on them so that the students can just listen. This way I try to engage and increase their understanding by going through longer proofs.

In graduate level mathematics courses, which are theoretical in nature, it is useful to discuss the patterns and ideas behind the proofs. I try to teach them the mentality of if you have this type of statement, these are the ways you can attempt to prove it. To be effective in teaching these types of courses, it is important to separate ideas from calculations and details. Keep zooming in and out so that the students do not get lost in the details and see the big picture, ideas and techniques that they can use for other proofs. It is helpful to begin with new concepts and give only the statements of the theorems and then go through the proofs and emphasize the ideas used. Also there are many concepts that the students need to keep up with in higher level mathematics. In these courses every Monday when their home-work is due, I give them a short ten minutes quiz to test only the concepts and give short problems from the previous week's lectures. These help the students keep up with the material, memorize and study the concepts and formulas as they do the exercises for homework. The quizzes are meant to be mainly on concepts and have short responses and few calculations. It is difficult to memorize many equations and theorems, so the short quizzes help students break it up and give them confidence in the material to prepare for exams.

To prepare lecture notes for graduate theoretical courses, it is difficult to do proofs without prior work and thinking on the problem, so it is important to balance between lecture notes and thinking for yourself in front of them. Have the ideas on how to do the proofs in mind so that you don't doubt and lose time and hence lose the interest of the students. Try to stimulate their thinking and walk them through the proofs. The goal is to connect with their minds, correct the confusions that they face and remember to be in control of your own thinking and think for yourself. Don't let their confusion make you doubt in what you are doing. Connect then disconnect and think for yourself. Have the strength in mind to help others think the right way in the material and at the same time be open and willing to listen to their reasoning and accept their correct ways to approach a problem to promote creativity. Empower students through creativity and deliver the material in a responsible way by showing the right way of thinking and reasoning through the problems.

In large classes, one has to be responsible to prepare every student for higher level classes and challenge him/her. To make sure that all students are following your leadership, you have to be organized and take the time to help those who don't have the right mindset for the material. I put a review day before each exam and after the exam I go over the questions with the class and share which parts of the exam most students did incorrectly and the common mistakes I found. I share with them the number of students that received a grade in the ninety range, eighty range and sometimes I share the number of students in the seventy range. It is important to recognize and have students look up to those who are doing well. An extra credit assignment also helps to give hope to students if their average grades are not as they had aimed for. A few weeks before the comprehensive final exam, I usually give an extra credit assignment with problems from the material throughout the course, so that it also helps the students to review and prepare for their final exam.

As for grading for large classes, I first make the solution key then grade one or two problems at a time for all students. After grading a few students' work, the solution gets to be on my mind and I can grade the rest more quickly. On the solution key, I mark the key steps that I should look for in the solution and scan the work they have shown to get these steps. If they did not get the key steps, then I look carefully in their solutions to see where they made a mistake. If they get the key steps correctly, then I am assured that all their work before it is correct and can just scan through them to make sure that they have shown sufficient work.

Most universities now use an online system for homework, which is usually developed by the textbook's publishing company. Students and faculty, especially in undergraduate courses, are adjusting to this new system and having it grade the homework automatically. From the instructor's perspective, the system can save him/her a lot of time; however, the main downside of online homework is that it tends to make students get accustomed to not showing all the steps in their homework as it is not being graded and no feedback on the work is provided. Also sometimes the students have the right answer and since they did not enter it in the exact format that the answer has been programmed, they get points taken off for being incorrect. I often get a feel for the level of the class and the students' maturity in mathematics through grading homework and by observing their handwriting and how neatly and completely they give their answers. These observations help me write the exams at the right level. During the past few years, I have switched to an online system for homework, as most other faculty, and grading the first exam has often become a surprise to me. For these systems, students tend to look at their notes while doing the homework and do well on them and since they get a quick response without comments and have to give the final answer without work, most do not spend as much time studying the material and thinking about it in depth as in the traditional setting. Giving short weekly quizzes and counting off points for not showing work, even if the final answer is correct, can compensate for these shortcomings. In mathematics the work and reasoning on how to get the answer tend to be more important than the final answer itself. Also it is helpful to advise students to print out the problems and do them on paper before filling the answers at the end and if there is any issue about being sure about an answer and getting it wrong, then they can send the instructor a picture of their work for that problem and the grade may be adjusted accordingly. To ensure that the students are keeping up with the concepts and techniques without looking at their notes, I randomly call on students and ask them questions in class. Thus, there are ways to adjust to online systems and not lose the values in the traditional ways of collecting and grading homework.

To ensure that all students are on the same page, it is helpful to post solutions to homework and exams on websites such as Blackboard or Canvas provided by the university. These websites also allow the faculty to send emails to students to communicate online. In addition, the students' average is calculated on these sites so that he/she is aware of his/her standing in the course throughout the semester. I personally record grades on a gradesheet on paper then enter them on these sites before each exam during the semester. A few weeks before the first exam, I give a thirty minute quiz so that they become familiar with the way I test, my grading style and types of questions I ask. To avoid discouraging students, it is important to make the exams challenging and stimulating, but not surprising. The questions need to be related and at a few levels higher than their homework to encourage them to study and think more through the material.

When making the exams, have every type of student in mind: those who are doing very well and those who are struggling in the course. There are some students who resist the mindset and no matter how clearly you teach the course or how easy you make the exam, they are likely not to do well. If you see this kind of gap between students, then go with the level of the majority of the students and form the exam so that it will be challenging and stimulating for the majority and not just the ones that are doing very well. Sometimes focusing on the strong students who learn readily can be comforting to the instructor; however, the instructor has to be for everyone in the classroom and especially for those who are meant to be there, to learn the material for the first time and thus are struggling. The student instructor relation should be that of compassion, respect and helping others. Helping students overcome challenges in mathematics can also help one recover from his/her own past challenges in mathematics and it is rewarding.

As for teaching a course virtually online, one of the main challenges is to get students to actively participate and feel part of the group. As is customary to use websites such as Zoom to meet virtually, I find that most become passive and do not show their image and keep their audio on mute in the meeting. To overcome this passiveness, I offer problem session zoom meetings twice a week for students to attend and solve problems. If a student does not attend either of these meetings in a given week, he/she gets points counted off from the homework due the following week and if the student attends both meetings, he/she receives a few points added to that homework grade. Points are like money to students and I find them helpful to encourage students to get involved in the classroom. Those who attend the meetings regularly and do problems correctly in class deserve to get extra points and they tend to do better on the exams than those who do not attend the meetings.

Before each zoom meeting, I type problems on a Word document and at the meeting I share the screen of that Word document and make it serve as the whiteboard as in a normal class setting. The equation feature in insert tab in Word can be used to type equations in the right format. Also having a touch screen laptop or an iPad and using a smart pen to write, can reduce the typing and be convenient. During these meetings, I randomly call on each student present and ask him/her to solve a problem on the Word file and I type his/her answer as is being said and pause to help out or correct the answer. This way other students listen and hear their classmates do problems and become more interested and engaged. Even if the students do not show their image, their names will appear on their screen and can be used to call on them. It is helpful to record each of these meetings and post them along with the Word document online on Blackboard or Canvas (where their grades are entered).

For online courses, to post lectures, I have found the website,

www.panopto.com

very useful. After the free-of-charge storage of five hours, it offers fifty hours storage for a monthly subscription. It allows one to upload and organize the video lectures in folders and share the links of the videos to students. Since there are many youtube videos available online on the material taught in the course, your lecture videos have to prove to be more understandable and connectable to the students. Making the videos long and monotone and not dynamic is less likely to be stimulating. I find that twenty to thirty minutes is sufficient for the duration of each video and is enough for me to explain a topic. It is important to have the students think with you as you go over the lectures. I write lecture notes and leave some problems without work and then hold my phone on top of the notes and record a video as I go over the material and solve the problems. This way they only see my hand doing the problems in front of them and I explain and go over the steps as I write them. Each video, along with its notes, is then posted online and zoom problem solving sessions are set to ensure that the students have followed and comprehended the material well.

Maintaining communication with the students is essential, especially in virtual classes. Try to respond to their emails not more than a day after. Even if you do not have the time at the moment, assuring them that you will answer their questions soon is more proper than having them think that their emails have been ignored. In online classes, email is the only way that the students can reach you and ask for help and not being responsive can turn them off. For virtual classes, I email the students every Monday, Wednesday, Friday at 9am., providing them the links for the zoom meeting, office hours for that day and give any information I need to share with the class, for example reminding them about an upcoming exam. I avoid emailing the class on any other time during the week. These emails offer them a routine to get used to and stay connected and up to date.

In addition to getting students involved, another main challenge in teaching virtually is to prevent students from cheating and using online resources. Websites such as, www.chegg.com allow students to post questions and within minutes receive worked out solutions. They can also type or take a picture of a question on an exam and get the answer on the site or as a text message sent to their phone if they download its app. The website does not provide the name of the individual who has posted the question and only keeps the question and answer on the site and it is a challenge to determine if someone used the source, even if his/her answer exactly matches that of the solution on the site. In addition, the website requires a monthly subscription and gives access to the solution manuals to almost all textbooks. In one class, some students were getting almost perfect scores on the homework, however, they were doing poorly on the exams and I discovered that they were copying solutions line by line from Chegg.com. Unfortunately, the use of online sources such as chegg is becoming more widely spread. Effective ways need to be introduced by universities to protect the value, meaning and integrity of education and prevent honest students from giving up and following those who are not disciplined to do the work themselves. Cheating should not be perceived as a "smart" way to move forward and it is the responsibility of the instructor to ensure that students are being evaluated fairly. Having students sign an honor pledge before a zoom or take home exam, helps to emphasize the seriousness of this issue. On the exam, if a student does not show sufficient work needed to get from one step to the next, gives the correct answer without the right reasoning or if exams from multiple students have the exact strange answer to a problem, then it is likely that an online source has been used. To avoid falsely accusing someone of cheating, when in doubt, the instructor can ask the student to meet in person and explain how he/she got the answer. If the student cheated, then it is very likely that he/she would not be able to explain the reasoning and it would prove to him/her that the penalty of points taken off from his/her grade is deserved for his/her dishonesty.

Mathematics is about thinking and developing one's mind and character. Many students understand this and willingly absorb mathematical concepts and enjoy learning and thinking through problems. However, because of the pressures of the society and the demands to be productive, more reasons need to be given to some students majoring in other fields. Discussing applications of the concepts often helps to motivate them. It is essential to bring in how the concepts and problems are used in modern times and companies. For example, as described in Chapter 7, Google uses million by million rows matrices to determine which website to list first and many areas in engineering model their problems using systems of differential equations similar to the systems students study in their textbooks. In cyber security, many papers such as the ones listed below use the system of equations that models a disease in a population to find information on the spread of a computer virus:

J. Roberto, C. Piqueira, V. Araujo (2009). A modified epidemiological model for computer viruses. *Applied Mathematics and Computation*. 213: 355–360.

O. Toutonji, S. Yoo, M. Park (2012). Stability analysis of VEISV propagation modeling for network worm attack. *Applied Mathematics Modelling*. 36: 2751–2761.

The system of differential equations used in the above papers are similar to the ones studied in a regular differential equations course and problems such as finding equilibrium points and determining their stability based on their Jacobian matrix are carried out in these papers. Showing modern research papers, such as these to the students, convinces them that the skills they are learning are being used in important fields such as cyber security and can be very motivating to them. The papers bring some light and purpose into what they study and better prepare the students for their future careers. They encourage the students to learn the concepts for the long term and tend to modernize the material taught, giving them the assurance that what they study will matter and be used.

Finding applications need not be entirely the responsibility of the instructor, and the students themselves can be asked to search online for them. In one class, I made an extra credit assignment before each exam for students to find a youtube video or an article on an application of a concept we had discussed in the class, write a half page paper describing it and then present it to the rest of the class. This assignment brought enthusiasm in the classroom, and it satisfied the students' need of purpose in the material. For large classrooms, the instructor can select some of the applications found by the students and share them with the rest of the class. In the beginning undergraduate courses in mathematics, applications are more difficult to find and the main purpose is to prepare students for higher level courses. Similar problems to what the students have studied and can do may be found in textbooks in upper levels and can be shared with them to give them confidence for their future studies. These problems can be solved in front of the students showing the book that they will likely study from in the upcoming semesters.

Students are there to learn from you and most will follow you if you gain their trust. It is essential to be honest with students and keep the mutual respect. Disrespecting and belittling students do not make one an effective leader in the classroom. Students should not be seen as a group to be lectured, but as individuals to be raised. It is essential to be patient and not say to each question asked, "this is elementary Calculus." Also avoid being an "I don't have time professor": when a student asks a question in his/her office, he/she says I don't have time and have research. When his/her collaborator in research contacts him/her, he/she says I don't have time, I have teaching. It is important to be organized and set time for each task and see the purpose in both teaching and research. When you are working on your classes and teaching, do not feel guilty that you should be doing research and when you are doing research, do not feel bad about neglecting your teaching duties. Each can have its time and your focus. Use the time frames you have during each week and throughout the semester and use them in the best possible way.

One of your goals in teaching is to make the concepts understandable and help the students overcome their fear of the level of the course. Sometimes when you help a student clear some past confusion in mathematics or solve a problem that has been bothering him/her for some time, is like making the world a little brighter for them and it is uplifting. Try to be in control of your own thinking and show confidence so that students will trust your leadership. Clarity of speech plays an important role in teaching. It is essential to speak clearly and be understood in class. When you are thinking deeply about a problem and your mind is occupied with your work and research, your speech tends to be affected. I find it helpful to read out loud. Reading a few pages of a novel or a piece in English literature out loud and understanding it as I read, helps me clear my speech and say words distinctly. If you are teaching in a large classroom and it is difficult to speak at the right volume, then a cordless voice amplifier is very useful to project your voice to everyone in the room. They come with an earpiece and a small device to put in the room to be heard clearly.

It is sometimes difficult to teach and have the mind to go backwards and remember past knowledge, especially after teaching it once and have to repeat it. It becomes even more difficult when you have to conduct research on the side and study and think about modern problems and think at much higher level in mathematics. It is important to balance and put blocks of time for teaching and research. This balance can be good for the mind to remind itself of past concepts and strengthen one's background. I view writing papers and research as my own homework and exams. Giving homework to students and helping them, make me more responsible in my own work and advancement in mathematics.

Always remember that you cannot raise others if you have not been risen yourself. Make sure that you do not fall behind in your work and research and that you are growing too and fulfilling your purpose. Research and being involved in the mathematics community help your own growth and should not be stopped, but one should balance to help others grow, as well.

A true leader is one who not only moves him/herself up, but also helps the advancement of others. Have in mind that it is more meaningful to help those in front of you than to write and help those whom you imagine will read your work. Try to help and connect to both groups. Remind yourself that writing a good paper and achieving a great result in research, do not mean that you know all mathematics, and teaching a lower level course and forgetting the concepts in that course do not mean that you know nothing in mathematics. Always improve in both research and teaching and aim to be the professor you always wished you had.

Math 1270

Ordinary Differential Equations

Section 1080

Spring (year)

Meeting Time: 12:00- 12:50 MWF
Meeting Place: ....
Instructor: Dr. Parisa Fatheddin
Office: .....
Office Hours: 11:00 – 12:00 am. MWF
Email: .....
Text: *Elementary Differential Equations and Boundary Value Problems* by W. Boyce and R. DiPrima

Course Prerequisites: Math 0280 (Introduction to Matrices and Linear Algebra) or Math 1180 (Linear Algebra) or Math 1185 (Honors Linear Algebra)

Course Description: This is a 3 hour credit course and provides the theory required for applications of mathematics in engineering, physics and chemistry. We will discuss how to solve first and second order differential equations and systems of equations. On each topic we will first focus on the theory then cover its applications. We will also cover Laplace transform and Fourier series as other tools in solving differential equations. General information on this course can be found at ....

Homework: There will be homework assignments from each section covered and the due date for each assignment is given on the tentative schedule. You may purchase a scientific calculator and use it for homework assignments; however, calculators will not be allowed on quizzes/exams or the final exam. Also you may work together on homework and email me questions or come by office hours. In addition, Math Tutorial Center is available and information on it can be found here: ....

Important Dates: Last day to drop/add: January 17th
Last day for extended drop period: January 24th
Last day for withdrawal forms: March 6th

FIGURE 1.1
Example of a Syllabus Page One.

Grades: Grades will be determined using the grading scale below. Your letter grade is a measure of your mastery of the course material and your fulfillment of course objectives. Letter grades are not assigned on the basis of a curve or the class average. All grades are recorded on Blackboard.

A	90 - 100
B+	85 - 89
B	80 - 84
C+	75 - 79
C	70 - 74
D+	65 - 69
D	60 - 64
F	0 - 59

Grade Distribution:

Homework and Quizzes	30%
Exams	40%
Final Exam	30%
Total	100%

Academic Honesty: All students are expected to attend each class session and take the quizzes and exams on the dates given. Your written assignments and examinations must be your own work. Academic Misconduct will not be tolerated. You may not submit in fulfillment of requirements in this course any work submitted, presented, or used by you in any other course. If you have any questions in this regard, please contact me.

Classroom Etiquette: Please be considerate of the instructor and those around you. Come to class on time and stay the entire period. Turn off and put away cell phones, laptops, iPods and beepers during class. Do not talk to classmates at inappropriate times. Refrain from reading or working on other coursework during class.

Disability Accommodations: Please let me know if you need any accommodation for seating, testing or other class procedures. Please see me after class or during my office hours to discuss appropriate modifications. You may contact the office of Disability Resources and Services located at ….

Quantitative Reasoning General Education Requirement: This course fulfills the Dietrich School of Arts and Sciences Quantitative Reasoning General Education Requirement (GER) given as follows,

All students are required to take and pass with a grade of C- or better at least one course in university level mathematics (other than trigonometry) for which algebra is a prerequisite or an approved course in statistics or mathematical or formal logic.

FIGURE 1.2
Example of a Syllabus Page Two.

Tentative Schedule				
Date	**Section**	**Topic**	**Homework**	**Due Date**
1/6 M	1.1 -1.3	Introduction, direction fields	Pg. 7: 4 Pg. 47: 2, 15a,c, 19a,c, 23 Pg. 39: 6c, 11c, 17	1/13 M
1/8 W	2.2	Separable Equations		
1/10 F	2.1	Linear Equations		
1/13 M	2.4	Difference between linear and nonlinear	Pg. 75: 8, 14 Pg. 144: 4, 8, 12, 18 Pg. 163: 10, 22, 23	1/22 W
1/15 W	3.1	Homogeneous second order equations		
1/17 F	3.3	Complex roots		
1/20 M	**Martin Luther King Holiday**			
1/22 W	3.4	Repeated Roots	Pg. 171: 6, 14, 16 Pg. 155: 2, 9, 13, 17	1/27 M
1/24 F	3.2	Linear Independence		
1/27 M	3.5	method of undetermined coefficients	Pg. 183: 4, 6, 14 Pg. 189: 7, 10, 15 Pg. 59: 2	2/3 M
1/29 W	3.6	Variation of parameters		
1/31 F	2.3	**Quiz**, Applications of first order equations		
2/3 M	2.3	Applications of first order equations	Pg. 59: 20, 24 Pg. 202: 5, 10, 18 Pg. 215: 6	2/10 M
2/5 W	3.7, 3.8	Applications of second order equations		
2/7 F	3.7, 3.8	Applications of second order equations		
2/10 M	4.1	nth order linear equations		
2/12 W	**Review**			
2/14 F	**Exam over Chapters One, Two and Three**			

FIGURE 1.3

Example of a Tentative Schedule.

2

Research and the Publication Process

From ancient times, we read that mathematicians used to find a solution to a problem and come out of their houses in excitement and share it with the public and then become famous and have their pictures on postal stamps and their names on streets and buildings. This is not the case in modern times, but the imagination of such an outcome can be inspiring for one to write a good paper. In modern times, the value of the paper is determined by the level of the journal it is accepted to. Research requires patience and determination and to write well, one has to read well and observe.

Writing a research paper is like writing a short story. The starting point is finding a problem that has not been done, then seeing a path (plot) on how to solve it (write it). A good writer first sets the main plot and major events in the story, then writes the details. It is about knowing the beginning and the end and then filling the middle and sometimes the end changes based on the details in the middle. Separating ideas from the details is the key, especially in mathematics, in which details are very important. In this chapter, the structure of research papers in mathematics and the publication process are described. An introduction to the computer program called Latex, which is commonly used to type papers in mathematics, is given later in Chapter 5.

A paper begins with a title, which in a few words states the problem that has been solved and studied in the paper. Underneath the title, is the name of the authors along with their institutions' names with a star indicating which author is the corresponding author, whom may be contacted for questions about the paper. In mathematics journals, the authors' names are given in alphabetic order; however, in most applied fields, the authors are listed in the order of the amount of contribution they made to the paper, with the first and second authors being the main authors of the paper. The main author in the journals in mathematics is referred to as the corresponding author.

The authors' email and mail addresses are either given on the front page or after the references at the end of the paper. In addition, if the research was supported by a grant, it is either noted in a footnote on the first page or mentioned in the Acknowledgement section. The keywords of the problem studied are given after the authors' names or in the footnote along with the addresses of the authors. Keywords might be the name of the equation studied and the name of the theorems proved. After keywords, one needs to look up the Mathematics Subject Classification (MSC) to find the number codes of the keywords. The classification can be found at

> https://mathscinet.ams.org/mathscinet/msc/msc2020.html

The keywords of the research area, theorem or equation name can be entered to find their number code. Then they need to be stated as primary or secondary, with primary being the keywords that are very much related to the result of the paper and secondary being the keywords that are somewhat related. A total of three or four codes is sufficient. For example, if the paper is about proving the Central limit theorem for the stochastic Navier-Stokes equation then,

> *Keywords*: Navier-Stokes equation, Central Limit Theorem, stochastic partial differential equations

> AMS subject classifications. Primary: 35Q30, 60H15; Secondary 60F05, 35Q35,
>
> where 35Q30 is the code for "Navier-Stokes equations,"
>
> 60H15 for "Stochastic partial differential equations,"
>
> 60F05 for "Central limit and other weak theorems,"
>
> and 35Q35 stands for "PDEs in connection with fluid mechanics".

Underneath the title and information on the authors, an abstract is required, which provides the reader a synopsis of the paper and states what has been achieved. This has to be a short paragraph and very concise and state only what has been accomplished in the article.

The first section of the paper is the introduction. Here you describe to the reader what problem you concentrated on and the challenges you faced. It starts with the importance of the problem, which in most cases are the applications to other fields and areas of study. One may find examples of such applications in other more applied papers and cite them. Then a short description on the problem needs to be given and a summary of previous works has to be noted with proper citations. To convince the reader of the importance of the result of your paper, state the novelty of your result and how it differs in setup or approach from other works in the literature. Afterwards, describe the main ideas and complications encountered to achieve the result. The last paragraph of the introduction should give an outline of the rest of the paper. It can start with "This paper is organized as follows," then state what has been done in each section.

In the introduction, avoid giving equations and most of it should be in words. If the paper studies a particular equation, then it is good to give that equation in the introduction with a short description on the variables in the equation. The background and complete information on the notations and problem have to be postponed for the second section. It is best to aim for the introduction to be two to two and a half pages, but not longer. The introduction is the first section that is read; thus, it has to attract the attention of the reader with its style and depth, and have him/her become interested in the paper.

The second section is devoted to the notations and main results. Here state all the spaces and notations that are used, the equations or problems that are studied and the basic definitions and concepts needed for the problem. Give a brief description and refer the reader to a few books that concentrate on this area of study. The paper should focus on your result, not on what is already known; however, explaining and giving some background provides the reader a good flow of the paper. Afterwards, state the main theorems that are proved in the paper without proofs. It can begin with "Here are the main results of the paper." These theorems are then proved in the rest of the article and their proofs can be broken into sections.

After the sections and the main content, the paper ends with Acknowledgements and then References. It is customary for the authors to state the grant number they were supported by in the Acknowledgements. It is best to write the Acknowledgements section after the paper has been reviewed by a referee, so that his/her comments and suggestions would be appreciated in this section. If one had conversations or help from others, without getting contributions from them in the paper, then their discussions can be acknowledged in the Acknowledgements section. It is appropriate to ask these individuals before mentioning their names in the article. Usually it is clear if someone wishes to be an author or does not mind giving ideas to others. Sometimes an idea from a person has a great impact on the result and entitles him/her to be named as an author. Therefore, one has to be careful and communicate with those who helped before the submission of the paper.

Each journal has its own format for references; however, in the journals in mathematics, it is common for a paper to be cited starting with the authors' names in alphabetic order with first initial and then their full last name. Afterwards, the title of the paper is given in normal font, with only the first letter of the first word in the title being capitalized and the rest of the words are kept in lower case. Then in Italics font the name of the journal is given using abbreviations that can be found online, for example, the California Institute of Technology's library at https://www.library.caltech.edu/journal-title-abbreviations.

One can click on the first letter of the journal on this site and then pushing keys control and F at the same time on the keyboard, can easily find the journal's name by typing some words of the journal's title.

In a reference, after the journal's name, the volume and issue number (if available) and then the page number range are provided. Depending on the journal, the year of publication is written after the authors' names at the beginning or after the volume number towards the end. Below are examples from different journals (*Journal of Functional Analysis, Probability Theory and Related Fields, Stochastic Analysis and Applications,* and *Communications in Partial Differential Equations,* respectively).

Korobenko, L., D. Maldonado, C. Rios (2015). From Sobolev inequality to doubling, *Proceedings of American Mathematical Society* 143: 4017–4028.

Rybarczyk, K. (2011). Sharp threshold functions for random intersection graphs via a coupling method. *The Electronic Journal of Combinatorics* 18(1): 36.

Kim, K. (2015). A BMO estimate for stochastic singular integral operators and its application to SPDEs. *Journal of Functional Analysis* 269(5): 1289–1309. DOI: 10.1016/j.jfa.2015.05.015.

Daus, E., M. Gualdani, N. Zamponi (2020). Long time behavior and weak-strong uniqueness for a nonlocal porous media equation. *Journal of Difference Equations.* 268(4): 1820–1839. DOI: 10.1016/j.jde.2019.09.029.

Most journals as in above, give the issue number in parenthesis after the volume number. Where to place a comma or a period should be observed and followed for each journal. To cite books, one begins with the authors' names in the alphabetic order with the first initials, then the title of the book is usually given in Italics font with the first letter of every word in the title capitalized. The name of the publishing company is provided afterwards, along with the location of its city and country. The following are examples from *Bernoulli* and *Communications in Partial Differential Equations,* respectively,

Ley, C., T. Verdebout (2017). *Modern Directional Statistics. Chapman & Hall/CRC Interdisciplinary Statistics Series.* Boca Raton, FL: CRC Press. MR 3752655

Feireisl, E., A. Novotný (2009). *Singular Limits in Thermodynamics of Viscous Fluids.* Basel: Birkhäuser.

Before submitting an article, it is expected that the references be in alphabetic order based on the first author of the book or article and the name of the journal, issue and volume, and page numbers be given. When writing a paper, I usually write the references in the form,

Nualart, D. and C. Rovira (2000). Large deviations for stochastic Volterra equations. *Bernoulli.* vol. 6, no. 2, 339–355.

Then after the submission to a journal and receiving their decision of acceptance of the paper with a revision, I change the references based on the format of that particular journal in the revision. Their style for references may be observed from a few most recently published papers in that journal.

As for writing the paper itself, first you have to decide on a problem that has not been solved and when you find it reasonable based on your time, search for more information about the problem and what has been done in the literature by using the Google Scholar and the database called

MathSciNet that is likely to be provided by your university. You may go to the website of your university's library and search MathSciNet under their website's research databases. MathSciNet has papers from every journal in mathematics and some from physics and other fields that have applications of mathematics. It offers a summary on each paper on its site and the PDF of the paper may be found by clicking on the link "Article". The following is an example showing how each article is presented on MathSciNet:

MR3668593 Saut, Jean-Claude; Wang, Chao; Xu, Li The Cauchy problem on large time for surface-waves-type Boussinesq systems II. SIAM J. Math. Anal. 49 (2017), no. 4, 2321–2386.

(Reviewer: Suheil A. Khuri) 35Q53 (35A01 35Q35)

Review | PDF | Clipboard | Journal | Article | 16 Citations

By clicking on the MR number at the beginning, a summary of the paper will be shown. After clicking on "Article", the journal's site will appear in another window and you may see "PDF" or "Download" on that site to click on to obtain the full text of the article and can save its PDF. Based on some of the journals' regulations, the paper might not be available through MathSciNet, and only the title and abstract will appear on the journal's site. In that case, Google Scholar may be used. Many authors post their articles on sites such as ResearchGate and the Cornell Library of Archives and if so, the link will appear on Google Scholar next to the title. If you cannot find the paper free of charge online, you may request it through your library by the service called the Interlibrary Loans. You may give the information on the title and journal of the paper or information on a book and they will send it to you electronically or as hard copies to be picked up at the library. It is a very helpful source.

The Cornell Library of Archives is a great website to check for papers and its address is as follows,

$$https://arxiv.org/$$

Most papers are first downloaded there, before they become published and are available for free of charge. The site is used as means of protection for the name of the individual or individuals who first solve a particular problem. A link to the paper on this website usually appears in Google Scholar beside the title of the paper. As an author, to download a paper on the Cornell library archives, you need someone in the mathematics research community to fill out a form that the website sends upon request, to verify that you are a reliable author. Once a paper is downloaded, it cannot be removed and even if it is replaced by a new paper, the previous versions would still show in the paper's history and can be viewed by others. Thus, one should download a paper when he/she is sure about its content and is about to submit it to a journal.

To search for papers, use keywords on the websites mentioned above and find other papers related to the problem. The Cornell library of archives needs to be checked to ensure that the problem has not been solved recently and that you cite the papers that have not yet appeared on MathSciNet. To cite the papers that have not been published, you may refer to them as preprint and give the archive number indicated on the archives' website for their paper. See below for examples from *Stochastic Analysis and Applications* and *Annals of Probability*, respectively,

Beck, L., Flandoli, F., Gubinelli, M., Maurelli, M. (2014). Stochastic ODEs and stochastic linear PDEs with critical drift: regularity, duality and uniqueness. *arXiv*: 1401–1530. https://arxiv.org/pdf/1401.1530.pdf

HILÁRIO, M., KIOUS, D. and TEIXEIRA, A. (2019). Random walk on the simple symmetric exclusion process. Preprint, Available at arXiv: 1906.03167.

MathsciNet is safe to rely on when searching for papers and simplifies the search since sites such as the Google Scholar search the keyword in journals from all fields of study, not just mathematics and can be overwhelming to search through. To search for papers, on MathSciNet you may choose "Anywhere" in the drop down menu in search and type the keyword such as the name of the theorem and then in another drop down menu type the equation's name or another keyword that relates to your problem. One can also choose "Title" in a drop down menu if only the papers having that keyword in their title, need to be found. "Anywhere" means the keyword is either in the title or in the paragraph summary of the paper provided by MathSciNet. By clicking on the MR number at the beginning of the title of each paper on MathSciNet, its summary will appear with the keyword highlighted.

It is not possible nor is it effective to find and know every paper and book that has ever been written on the subject. Scan through what you find and pick the right papers and books that are related and have the potential to help you. The papers in higher than average journals, give a good summary of what has been done in the literature and you can use these references to find other papers related and form your own list of references.

To turn the research process into stages, I usually start by going through one round of search, in which based on the papers' titles and abstracts, I save the ones that might be related on my computer. Then I look at each one and rank them as 1 for being very much related, 2 as somewhat useful and 3 to keep just for studying and may be used as references in the introduction. Each paper can be saved as for example, 1 Wu, where 1 is the rank and Wu is the name of the first author. The computer will automatically order the papers saved, numerically, if they are saved in one folder.

For round two of the search, I print the pages of references at the end of each paper in ranks 1 and 2 from round one and then read the introductions and circle the papers they refer to and then find those papers. It is likely that you will find some papers that are frequently referred to in different articles and seem to be widely read in the area. To be complete, it is good to refer to these papers in your paper. The number of citations of a paper or book is how many times the paper or book has been referenced by other authors. It is acceptable to cite your own previous papers if they are related to the paper you are currently working on. MathSciNet offers the number of citations for each paper underneath the title and description of the journal as can be seen in the example above. I think two rounds are sufficient to get one started on the problem and form the introduction. I prefer to write a draft of the introduction of the paper first, in order to study the papers and have an overview of what has been done before doing the problem.

It is best to separate and mark the papers that are well-written and deep as your model papers to be studied in detail and be the focus of your study. You will likely find papers that have your style of writing and those that are written in the language you can connect to. Make sure to study the papers that have been widely referenced and include them in the first rank. Focus on one model paper at a time and get the background you need to learn and understand the domain of the problem. Your paper has to be based on facts and observations of ideas from previously published papers in the literature. Keep reading and studying until you see it and find a path. The more you read and get into the mindset, the better you will see and understand what needs to be done. After some time studying the papers and material and looking for patterns, your mind gets trained for the mindset and you can then learn the concepts in the research area more quickly.

For every paper, it is important to set a satisfaction point, since no matter how much you add and make perfect, the paper can always become longer and more in depth and improve more. Try to get the paper to the best state it can be, based on the time you have and then submit it to a journal. You can submit a paper to only one journal at a time; thus, selecting a related journal is essential. It is better not to risk too much on the level of the journal. On the homepage of MathSciNet, where you can search for articles, there is a "Journals" tab on top. After clicking on it, you may enter the name of your area of study, for example "finance" or "partial differential equations" and all the journals having those words in their titles will appear. By observing some articles in each journal, you will find some that seem suitable for your article. Depending on the result, some journals will seem safer than others and it takes time to find the ones that match your style of writing and research area. Many times, choosing a journal from those papers you have studied to achieve your result that match your style is the best approach. No journal guarantees acceptance and one needs to be prepared for possible rejection but can reduce the likelihood of rejection by observing similar papers that have been published in the journal and make a careful comparison before submitting.

After you decide on a journal and submit your paper, it takes a short period of time, usually less than two weeks, for the editor to approve the paper to be reviewed by the journal. After this approval, you will receive an assigned number for your paper. The associate editor, then sends the paper to one or two individuals, referred to as referees, to decide if the paper is suitable for the journal. Depending on the journal and the referees, the review process typically takes two months to six months or even in some cases a year or more. Once the referees have completed their evaluations, they provide the associate editor their report. The author will not be notified on who the referees were and some journals are double-blind, meaning the referees are also not aware of the authors' names of the paper in order to give a more fair judgement.

The associate editor and the referees have to agree on the decision. If the referee rejects the paper, then it is very likely that the editor will also reject it and another journal should be considered. If the referees offer reports that share comments and corrections and state that after a revision incorporating their suggested changes, the paper is suitable for publication, then the paper has a very good chance of getting accepted. One has to carefully answer each comment and make the changes that are stated, accordingly. Be sure to take care of every comment and refer to each by its number. Avoid changing other parts of the paper, other than small typos, if no comment is made on them, unless it is essential for the results and in that case, explain and state the change clearly in the response to the referee report. Most authors use blue or red text color in latex to indicate where changes were made in the revision.

Most journals allow a period of two weeks for the revision based on referee report. It is best to try to respond in about a week as the paper is fresh on the referee's mind. To write a response to the report use Latex as in the paper and refer to each comment by its number and correct what the referee has pointed out. One needs to give detailed explanation of the questions asked. In your response, it is proper to begin with expressing appreciation for the referee's time. Write the response in a respectful and professional manner and in the Acknowledgement section of the paper, express your gratitude to the referees' time and help in improving the paper. In the Acknowledgments section, it is more appropriate to refer to the referee as "he/she", instead of "he" to be inclusive and respectful of the female researchers and referees in your field. In the revision, in order to try to help the associate editor, it is favorable to change the format of the paper and especially the references to match those of the journal by finding and looking at its recently published papers.

In the case that the paper gets rejected, keep in mind that sometimes the referee does not see the result of the paper from your perspective and might not be in the exact area of study to observe its importance. Therefore, selecting the right journal can help save time. Before submitting a paper, it is best to prepare yourself for possible rejection and have a plan of where to send it next and what can be done to improve the paper. Take criticisms objectively and professionally and respect the referee and editor's decision, then improve the style and writing of the paper before submitting it to another journal. The referees usually point out mistakes that caused the rejection and correcting them help increase the value of the paper.

After putting so much time and effort on the paper and waiting for months for the decision, getting a rejection is not pleasant. It is like a tree, from which branches have been trimmed and is not left to grow and reach the sun for more light. You have to grow again and not let negativity dry your tree of knowledge and growth in mathematics community. When you get a rejection, it does not mean that the mathematics research community has rejected you and if you get an acceptance, you have to remind yourself that you still have more papers to write and have not yet reached the sun to stop growing. After getting trimmed, some trees dry out and the trees that grow back, their new leaves and branches tend to be stronger than before. The goal is to improve and grow. The taller your tree of knowledge becomes, the better you will see from above and you will be surrounded by more light. Try to improve your techniques of writing and write stronger with more depth and insight and note that a good paper requires many revisions.

When you write a paper, imagine the positive outcome. That an author you know and admire will be the referee and will appreciate your work and understand your way of reasoning. Imagine that the mathematicians you admire from model papers will put the time to read it carefully and give you positive and useful comments. That they will be interested in understanding your idea and what your paper offers to the research area. To respect their time, make sure that the paper is in the best possible state before submitting and that it has no typos, nor language and mathematical mistakes.

If you yourself become a referee to review a paper, first read the title, abstract and introduction and see if you can responsibly take on the task based on your research area, knowledge and time. If you decide to take this responsibility, then read the paper carefully and act as the referee you like to have for your own papers. First read the introduction and statement of main results to gain an overview of the article and understand the main ideas. Write a summary in words in your notes and think about ideas as you read. After checking the accuracy of the main ideas and approach, make sure that the results are novel and that the important papers related to the results have been cited. Then go through the details and estimates and point out mistakes if any. Read piece by piece to find if it is all correct and reasonable. The key is to be responsible and fair in the evaluation. Try to take notes and understand the paper in depth. It is very likely that you will learn some technique or idea from the paper to add to your own understanding and knowledge in the subject. When reading, the order of your priorities in evaluation should be the level of ideas and length, accuracy, then style and organization.

To prepare the referee report, start the page with "Referee Report for 'the title of the paper' " and if the authors' names are shown to you, then state "by 'authors' names' " as part of the title. The report then begins with a summary of what has been achieved in the paper and the novelty of the result. The comments on the paper as a whole on style and organization are given in this paragraph and it is stated if the paper is suitable for publication in the journal. Expressions such as "it is interesting", "well-written", "it is novel", "organized", "not suitable", or "good contribution to the literature" can be used. You have to prove to the authors that you have put the time to read their paper in depth by providing them detailed comments to improve the paper. Number the comments and indicate where in the paper changes need to be made by stating the page numbers. Some reports also provide the line number within the page where the paper needs to be modified.

In the review process, try to help others as much as possible. They have put the time and effort into the paper and are also waiting for the right feedback. Read with an open mind and be inclined to accept if the paper has the level of ideas and depth that match those of the journal. The associate editor has already approved the paper for the review process and has observed its potential for acceptance to the journal. However, if you believe that one of the key ideas is incorrect and that correcting it would take more than two weeks then rejecting the paper would be reasonable. Also if you find repetition of the authors' previous works in the paper or their use of other papers' results without the addition of their own creativity and innovation, then rejection would be appropriate. The level of the paper as a whole should match that of the journal. Be professional in your report and avoid writing in a negative tone. Even if you have decided to reject the paper, it is good to give some good comments on the paper and ways to improve.

Try to decide early in the review process if the paper should be rejected or it can be accepted after a revision. This way if the paper is sure to be rejected, the authors can be notified early on and it saves them time to submit to another journal. The worst a referee can do to authors is to wait more than six months and then reject the paper without giving comments on improvements and only write one sentence indicating that it is not suitable for the journal. If the paper's main ideas are accurate and the paper is at a suitable level, then one may take the time to provide the report on comments for revision and aim to complete it by the deadline requested by the journal. You are grading the paper, but should not only be focused on finding mistakes and giving negative comments in the report, but it is better to also appreciate and understand the results that are accurate and well-written.

One can think of the R in research standing for the word Risk. If others knew how to do the problem then it would not be a research problem. Research is about contributing to the literature by bringing in new results and adding to the understanding of a research area. One should value his/her observations and believe in the problem and way of proving it. The risks taken need to be to a limit. It is important to be determined, but one has to also avoid getting stuck on one problem for more than the allowable time, based on his/her position in the university. Time is an important factor and the balance in learning and building a foundation with achieving publishable results is the key. Sometimes applying a recently proven technique to other problems or solving mathematical problems in applied fields seem more safe to undertake for the first few papers.

In the process of writing, doubt to the point of being careful and being perfect, but not to the point of blocking your thinking and creativity. Papers require creativity along with clarity and strength of mind to think abstractly and then to bring ideas down to write them in an organized way to be understood. Each paper should be a good piece of work and like a painting in which if the details are painted well, the painting as a whole becomes spectacular and intriguing, but if one detail is not painted correctly, it will show as a viewer looks at the painting carefully. Every detail and idea has to be correct. One needs to be brave to come up with new ideas and write with confidence, but also be responsible, since the results will likely be used by other individuals to study or in other papers for new results or in industry to be applied in the real world.

3

Conferences and Collaborations

One of the great satisfactions of solving a problem is to share it with other mathematicians and thus feel part of the group. Attending and giving talks at conferences are very important for one's research development. Writing requires inspiration and conferences can be a great source of inspiration. It is important to look up to those who have achieved more and listen to their talks. I find conferences and attending talks as essential fuel for my research. Seeing other mathematicians working on the same type of problem as mine, gives me great motivation. Also meeting the authors, whose work I have been studying and admiring, is uplifting. Hearing talks in the same area is likely to make the problem you are working on seem less difficult and the result more attainable.

Mathematics has many branches and an extensive amount of research has been done in the many areas of each branch. Thus, it is impossible to know everything and one can effectively know few areas well. The core mindset for all areas is the same and as a graduate student in mathematics you have the core mindset to be able to follow at least some pieces in every talk. It is good for the mind to listen to research talks and be able to pick up different ideas and ways of thinking. One should aim to catch up in other areas as much as possible in addition to being responsible in knowing his/her own area of research well.

Before attending a research conference, in order to better follow the talks, it is helpful to at least read and write the titles and abstracts of the talks a day or more before, find the definition of the keywords and search and gain some background knowledge on the topic being discussed. It is disrespectful to the speaker to attend a talk that you have no prior knowledge on or are not interested in learning about the topic and thus will likely tend to use the time and speaker's energy and mental stimulation to think about your own problem. When listening to talks, try not to tune off and say this is not in my research area. Try to be open to new research problems, in addition to those within your research area. It is good for the mind to get exposed to other ideas and ways of thinking. Try to follow and pick up concepts as much as possible.

To help the progress of your own research problem, however, it is preferable to attend the talks that are closely related to your research area or those that are on a topic that you have some knowledge on and want to know more about. One cannot expect to learn an area of study by attending talks, since

DOI: 10.1201/9781003299073-3

most speakers focus on their results and assume that the audience has the needed background. Even though research talks are not expected to have a teaching component, they can, however, offer an exposure to an area of study and way of thinking. A good talk starts with some definitions and explanation of the main concepts to get everyone on the same page and offers understanding and depth during the talk. The main purpose of the talk should not be to impress and intentionally confuse the audience so to make the results seem out of reach and noble. The main purpose of the talk is to discuss the results and have the audience be absorbed by the depth and knowledge provided during the talk and let them understand and follow the reasoning used in the results.

If a talk is in the same branch of mathematics as yours, it helps to look at the paper being discussed before attending the talk. Typically speakers make the title of their talk similar to that of their recent paper and in most cases their paper may be found on the archive:

www.https://arxiv.org

One can search the speaker's name and find the paper among his/her latest downloads on the archive. Reading the introduction of the paper and finding and writing the ideas and main results, can greatly enhance your understanding of the talk. Before the talk, as you look into the paper, think of questions that come to your mind. For example, why the author chose this method and not the other method that you have seen in the literature, what were the complications in achieving the results, can the results be implemented on other similar equations, why this particular equation was chosen and not the other equations. Also questions about the concepts and ideas used. Writing down these questions and having them in the back of your mind as you listen to the talk, help you listen actively and follow the talk. At the end of the talk, if the questions you had were not addressed, you may ask one or two of them. Speakers like to have some discussions at the end and will likely welcome your questions. It is best to write down questions and ask at the end of the talk, since research talks are timed. Try to avoid questions on basic concepts and those that might seem elementary to someone who has studied the area. However, if a question raises your interest on the topic then ask and do not let others who have studied the topic for some years make you feel intimidated and apart from the group. In conferences, achieving good results and contributing to the literature give you the right ticket to be admitted to the conference and be part of the group. There is always gaps between the levels of knowledge in the audience, but the main goal of the conference is to make sure that everyone gets the most out of it.

Sometimes even if you read all the paper before attending a talk, you may not understand what is being said during the talk. In that case, write the terms you hear. Keep your mind active by taking notes. It is best to follow and understand talks and not just listen but if they are not at the right level of what you know, then writing down terms as you hear them will make those terms seem less foreign when you see them later in papers or hear them again in other talks. After the talk, you may also search the terms and find information on them so to prepare for future talks on the topic. It is important to have some background on each branch in your field. It is also essential to hear the correct pronunciation of the terms. Hearing them and seeing how they are discussed, will help your studying of papers more active.

As for the first conferences to consider to attend, I recommend the American Mathematical Society (AMS) and the Mathematical Association of America (MAA), which are two well-established conferences in mathematics in the United States. Both have many sectional meetings every year and one annual joint meeting of AMS and MAA together in early January. The calendar of the times and locations of the meetings can be found on their websites,

https://www.ams.org/home/page and https://www.maa.org/

Research conferences organized by the Society for Industrial and Applied Mathematics (SIAM) are also very informative and helpful to attend and participate in. The calendar of their events can be found on their website:

https://www.siam.org/

SIAM often holds panels and sessions on the many uses of mathematics in industry and one can get introduced to careers in companies that apply research in mathematics and hire individuals with a Ph.D. in mathematics.

On the AMS and MAA sites, you may click on "sectional" meetings and find the upcoming sectional conferences in each. On the AMS website, you can choose a meeting and in its program schedule can find special sessions. The title of each session indicates the topic of research it plans to focus on. For example, "Special Session on Stochastic Partial Differential Equations." In any upcomig sectional meeting, it is very likely that you will find a special session with a topic that is closely related to your research area or field. For your first research conference, it is best to only attend and listen and observe the talks. Afterwards, if you are close to finishing your paper and are ready to submit it to a journal, then you may consider giving a talk.

Giving a talk on a paper as you are at the final stages of writing, is very important to settle your thoughts on the problem. As you prepare the slides and organize the material in a way to be understandable, you zoom out of the problem and see the big picture and review and remember the process you took to attain the results. A talk can be seen as an examination on your paper and on what was done. Your contribution to the research community should not be limited to writing papers, but also being actively involved by giving good talks.

On the AMS website, after choosing a meeting, you may submit an abstract before its deadline and wait and see if they have available time to include your talk. You may also contact the organizer of the particular session that fits your research topic and share your paper and express your interest in giving a talk. If you are new to the research area and to its community, then submitting an abstract is the more proper choice. Most organizers accept and include talks that are submitted to their session. If they have never heard you give a talk, they will likely schedule your talk toward the end of the session, but that is sufficient for your first talks. After participating in conferences more regularly, you will see individuals that you can recognize and connect to and can contact them if they become an organizer. Then later when you find a group of researchers that share your interest in your research area, you yourself can organize a special session.

Each speaker, even if invited by the organizers, has to submit a title and abstract for his/her talk. The title should be concise as in a paper and the abstract should not exceed a paragraph. It is best to avoid equations in the abstract. If you deeply understand your paper, you can summarize its results in two or three sentences. The abstract should give what problems will be discussed and the main approach that was taken for each problem. At the end of the abstract, mention the names of those who also contributed to the paper by stating for example, "This is joint work with..."

In most conferences, the duration of each talk is about 20 minutes with an addition of 10 minutes for questions and comments. One can typically combine the two slots of time and aim to finish in 30 minutes. The invited talks and those given by well-established mathematicians are typically 50 minutes. If you are planning to give a talk, have in mind that since the people attending the conference have devoted the time of the conference to actively think and focus on their research, it is not difficult to convey the information on your problem and results in a 30 minutes time slot. It is likely that most will actively listen and follow your talk if delivered well.

The conventional structure of a research talk in mathematics is described as follows. To prepare the slides for the presentation, the package called Beamer is used in Latex and is explained in Chapter 5. The presentation begins with the title of the problem and the name of the speaker and then the date and location of the talk. One should also include the names of other authors of the paper by the statement "Joint work with...". At the beginning of the talk, state the problem/problems. You do not want to only focus on the details and at the end, some in the audience may not be clear on the actual problem you worked on. The big picture should be given clearly at the beginning. Many give an outline of their talk on the slide after the title slide. To motivate, it is good to start with an introduction to some concepts and definitions on the area of study. Some past results and explanation on what has already been done on the topic in the literature, can help gain the interest of the audience. Discussing applications of the area of study and showing related pictures can bring some color to the presentation and emphasize the importance of the contribution. Some also show pictures and videos of their numerical simulations toward the end of their talk.

When stating the theorems and main contributions, give the name of the authors with dash or comma in between and the initial of your last name then the year of publication. For example,

Theorem 2 (B. Abel, H. and D. Jones 2019)

where the speaker's last name starts with H. In the talk, mainly focus on the ideas and avoid showing calculations and details. Do not put too many words or equations on one slide. As in the paper, you want to have a structure in what you present and not lose the audience.

During the talk act as a reporter by reporting the problem you have worked on and on how you solved it, convincing the audience of the reasoning in your proofs. Show confidence and pride in what you have achieved and give the talk as if the authors of the model papers you have studied for your paper, are in front of you and are actively listening and are interested in knowing the new idea and result you have contributed to the area. Share the results clearly and explain the facts and reasoning used in the paper to support them. If questions are asked during the talk, then see it as a good sign and answer them to the best of your knowledge and if you do not know the answer to them, then it is ok to admit it. You have not studied this area of research for many years and it is acceptable not to know some aspects of it. It is best to welcome conversations as much as possible and also keep a note on the time that you are given for the talk.

Engage the audience and in a conversational tone discuss and talk about the problem to get them interested and involved. View the individuals in the audience as future collaborators and referees for your papers. Aim to have them follow you and not just sit and listen and clap at the end. Show the confidence and empowerment you gained by achieving the results. The only person who can follow and understand your talk should not be the one who has worked on the particular problem him/herself. Slowly get the audience into the mindset of the problem and walk them through the results of the paper. There is no need to explain the basics, but it helps to start with some background and explanation on the topic to get the audience on the same mindset. The beginning of your talk may include information from the introduction of your paper. It is good to discuss similar results in other papers and state their differences with your article. Afterwards, state the main results and then explain how you obtained them and give the main ideas in overcoming obstacles. You want to respect the time of the individuals in the audience and help them get a feel for your research area and the problem you worked on and have them learn and pick up some useful information from your talk.

Not everyone in the audience will be in the same research area and might not be actively listening. In the talk, it is best to aim to engage as many as possible and be mentally stimulating and show interest and want everyone to follow and understand. At the beginning, if you do not see the majority to be interested in your talk, do not rush through the slides and lose your ground. Try to go at the right pace as you previously planned it. Make slides organized and the information to flow the right way for most to understand. Before the talk, I usually print out four slides per page and write small notes underneath the slides to remind myself on what needs to be said for each one. Before the talk, I review the material and model papers and get into the mindset and domain of the problem. One has to review as much as possible and know the domain of the problem well before giving a talk. After preparing the slides, it helps to write two or three pages in regular language and essay format without writing equations on what needs to be said in the talk. This helps put all the information that can be said in one place and reading it before the talk helps bring back the material to your mind.

Typically talks are grouped by subject area and there is a good chance that those in the particular area of your research will be in the audience and they might even be assigned to read your paper and decide on its acceptance. It is your chance to convince others of your reasoning in the proof and explain it well. Show the understanding, thinking and learning that went through the research to write the paper and achieve the results. The talk should be in a way to invite others to work in the same area and maybe become your collaborators in a related problem and paper in the future.

After attending some conferences and becoming more developed in your research, it is good to consider organizing a special session yourself. I find organizing special sessions for AMS meetings enjoyable. It is nice to invite and gather researchers in the same area and try to provide the right setting for them to discuss results with one another and understand each other's accomplishments. One needs to submit a proposal to the main organizer of the sectional meeting before its due date indicated on the website. In the proposal, begin with the title being AMS proposal for which sectional AMS meeting and give the date and name of the conference. Then the title you propose for your special session should be given. Also the first two digits of the Mathematics Subject Classification (MSC) of the research area has to be provided. Afterwards, give your name, email and contact information and name of your affiliated institution. A short description of the session with why the research topic is important should follow, giving a brief explanation of the topic and its applications and importance. Discuss how you plan to carry out the session and assure the organizer of your own achievements in the area of study. Then a potential list of speakers needs to be provided. This list need not be the final list that is used to send out invitations and may change. It gives the main organizer the assurance that you have a list of speakers ready to invite. I usually give about twenty speakers with their names, along with the name of their institutions and their email addresses. This list also helps me to plan for the session and invitations.

As for attending any talk, before your session, have a good idea on each speaker's talk based on his/her abstract and the paper you find on the archive and actively listen to each talk. Have in depth questions to ask to ensure there will be some discussion at the end if no one else in the audience asks a question. As the organizer you have the opportunity to provide an atmosphere in which speakers can freely discuss their research and have conversations during talks without being under time pressure. The rule is to give each speaker up to 30 minutes, however, this is not a presidential debate and if they go over that time period and there is a good discussion happening, then you can give them a few more minutes and adjust the time frames for the next speakers. Your goal is to offer the setting for individuals to understand each other's research and through discussions, help each other in the problems and think of ways to apply and extend the results. If you keep looking at your watch and making the speakers be so careful about not going over their time limit, you would not provide that kind of atmosphere.

After you attend and become accustomed to the national conferences and you have made good progress in your research area, you can then try to attend conferences devoted only to your specific field of study in mathematics. There you are likely to see many of the authors whose papers or books you have studied and can find those whose research problem closely matches yours. Sometimes when you are stuck on a problem, attending talks and discussing the problem with others help your mind pick up other perspectives and thus be stimulated and get the motivation needed to overcome the barrier. Also during the conference, you can often find researchers who are willing to have discussions on the problem you are working on and help out.

Because of the challenges in writing a research paper, it is essential to collaborate with others. Solving a research problem in mathematics is a difficult task and some help from those in the research community can really make a difference. Collaboration can make research very active and mentally stimulating. Some collaborators act as mentors since they are at a much higher level in research and some collaborators are at the same level and have the same ideas and interests and some can be in different fields of study or area and make the paper interdisciplinary. In all kinds of collaborations, research can be stimulating and having others share your interest in your problem, can give great motivation and enthusiasm in conducting research and solving the problem. Since not many individuals know the concepts in the area of study of your research, it is fulfilling to discuss them with collaborators and it gives assurance to one's own in depth understanding in the subject area.

The key to a productive collaboration is good communication and it is important to be responsive. It is essential that each individual contributes his/her share and follows the required teamwork to make it a success. I usually set deadlines for myself and let the collaborators know when I plan to accomplish a part of the paper. This way they can be assured that I am working on a specific aspect of the problem and they can focus on their own section and contribution. Avoid postponing your work on the paper and not delivering what you promised. The worst a collaborator can do, is to keep promising and give false hope to other members of the group and delay their work. It is good to be honest and promise what you can do in the time you are given.

In every collaboration, one or two individuals are the driving force of the work and motivate others to contribute and they themselves contribute the most to the paper. Try to aim to be the driving force in every collaboration. There are different kinds of collaborators in mathematics. Some contact you frequently and come to your office on a regular basis and some take the problem and you do not hear back from them in a month or two. Those in the second group tend to wait until the inspiration strikes, then they begin writing. Some, especially those who have previously published in highly respected journals, can see through the paper and in a short period of time, contribute the missing piece to the result that helps finish the paper. You have to adapt to each person's working style and do your best to make the collaboration successful.

Some collaborations work between individuals in different fields, where each person contributes to the problem based on his/her expertise and at the end, the paper increases in value by being an interdisciplinary article. For example, one can show the theoretical proof of the results and another person can verify the results by using computer simulations. In the case of an interdisciplinary paper, it is important to make sure that the different contributions have the same style and similar way of expression to make the paper appear in one piece and all ideas be accessible to the reader to follow.

Try to avoid collaboration in what I call the "ping-pong" style. That is the collaboration, in which you send the paper to the other person hoping he/she will add to it and he/she sends it back to you after some time, hoping you will add to it and figure it out and the paper is sent back and forth many times without much progress. It is important to have patience and motivate the collaborators by contributing to the paper yourself first, then sending it to others. One has to be positive and active and prove to others in the group the interest and dedication to the work by bringing accurate and well-written results. There is no need to rush in your share of results only to keep the communication active. The goal is to make contributions that others can rely on in a timely fashion.

In collaborations sometimes you work with someone, and sometimes you work for someone. As a graduate student and later as a postdoc, you are mainly working for your mentors and need to adapt to their research area and style of writing. It is very important to learn and observe from those who have made many contributions. Respecting and upholding those who are more experienced are key parts of the culture of mathematics. They can see through the problems more clearly and give you ideas to achieve the results. They can also have a better overview and help you think of future problems to take on and extend your results. Their insights rightly deserve to be greatly appreciated and one needs to note that their time to advise and guide as a mentor counts as service to the department and most do not get compensated by increase in pay. Respect and appreciation are what they expect from their students and most view advising as a way to help the next generation move up in research.

Because of the increased amounts of pressure and number of expectations in academia, you can not expect your advisor or mentor to be the only source in getting caught up in the research process. Each discussion and meeting with your mentor should help you get the ideas needed to make good progress in your work. Being able to find the right idea is essential and is what you have to learn to develop to become a stronger researcher. One needs to acquire the ability to learn so much material and be able to connect the dots the right way. Advisors usually try to push you to become more independent and think for yourself and become more confident on your own ideas and creativity. Going to conferences and seeing other well-established mathematicians to look up to help build that confidence and independence.

An ideal advisor can be viewed as a father or a mother figure that guides you through the research process, helps you develop your ideas and is there to help out if you are facing a block in your problem. It is best to aim for this ideal relation, but also have in mind that the expectations of faculty research have evolved from the past and most need to prioritize their own research and projects. Their time of advising should be valued and by bringing good results and knowing the material well to be able to effectively talk about the problem, would gain the interest of the mentor and increase his/her involvement in your problem. As a graduate student, one should keep track of the time he/she takes to complete the research projects and balance it with the time necessary to become prepared to conduct research independently before graduating.

It is enjoyable to discuss research with others and work together on a problem, but it should be seen as a serious task to be completed. Each individual has to make some contribution to the paper to be included as an author. If a person offers an idea that becomes a key ingredient of the paper, then he/she needs to be included as one of the authors. In some cases, especially in conferences where many like to discuss problems, some individuals might give ideas and offer help in the paper without wanting to be part of it. It is important to contact them and see if they want to be named as an author and even if they do not, their help has to be acknowledged in the Acknowledgement section at the end of the paper.

In mathematics the number of authors on a paper is noted by other researchers. In applied fields, the authors are listed by the order of their contributions with the first author being the main author and it is not unusual to have more than five authors for one paper. Many times, the name of every person in a laboratory is listed as an author even if he/she was not part of the project. This is not viewed as ethical in the academic setting and some contribution to the content and results of the article has to be made to entitle one to be listed as an author. Every university has guidelines on the principles and practices in the university research. For example, the Authorship Guidelines of the Ohio State University states the following paragraph on what is unacceptable in regards to authorship.

"The following authorship practices are not in line with the criteria established for authorship and the values of Ohio State and should not be allowed:

a. Guest Authorship – the practice of assigning authorship to someone who has not participated in the work, simply to honor that person or to provide additional credibility to the submission based on the status or standing of the guest author.

b. Gift Authorship – the practice of assigning authorship to someone who has not participated in the work, to reward him/her or provide an unearned benefit.

c. Ghost Authorship – the practice of not providing named credit to individuals who have made substantial contributions to the work or in the writing of the manuscript. This often is seen with the use of professional writers who are not credited or acknowledged. Writing activities alone, such as writing assistance, technical editing, language editing, and proofreading, without other contributions may not qualify for authorship, but should be acknowledged."

Ghost writers are those who are paid from the named author to write a paper and make a major contribution without being recognized as an author and their help is not acknowledged. Unfortunately, because of the increased, somewhat unreasonable expectations on the number of publications from the faculty, the above practices, especially ghost writing have become more widespread. Also the "Big Shot" culture and relying on connections and friendships between the faculty, have contributed to the increase in the first two of the above practices. One has to go through the system, but also keep in mind the ethics that are involved to be truly successful.

As discussed in this chapter, attending and giving talks at conferences and collaborating with others play key roles in one's development in research. In conferences, one devotes two or three days purely to research and thinking about problems and can gain many new ideas. Being with other researchers and discussing ideas and concepts are essential for the in depth learning and understanding of a research area. To promote research and collaboration, most universities offer weekly research seminars and colloquiums and invite researchers from other universities to visit the university and give a talk. The seminars also offer the faculty and graduate students within the university the opportunity to present their research and new results. Before giving a talk at a conference, one should take advantage of this opportunity and present at a seminar and ask some of the faculty for comments and suggestions on how to improve the slides or the presentation.

Seminars are one hour long and provide sufficient time to discuss the results during the presentation with other faculty than one's advisor and thus one becomes more confident about the reasoning used in the results. If the problem and method to solve it are explained and discussed well, then the faculty in the audience would be acquainted with the results and later would likely accept to be part of the individual's Ph.D. committee, when he/she has to defend his/her dissertation.

4

Transition to Reading and Writing Research Papers

The process of writing a research paper in mathematics is like the process of building a house. First one needs to build the foundation (the problem) and sketch an overall structure (the plan to solve it), then place in it windows and doors (theorems and calculations) and afterwards give it an interior design (making the language of the paper to flow) in order to make it suitable for the occupant (reader). Sometimes it is easier to work on a rundown house and improve it as long as the foundation (problem) is good and other times it is easier to start building a house from ground up.

At the beginning, it takes some time and effort to adjust to the level of articles and learn how to conduct research. In typical graduate courses in mathematics, most proofs that one studies and is asked to formulate are at paragraph's length. It becomes a great challenge to then work on a problem that requires a proof of length twenty to thirty pages. You have to catch up and speed up your learning to know the problem and area of study and understand the research process. It is like driving and all of a sudden entering a highway. In the early stages, because of the immense amount of information that needs to be learned, it might also seem like driving in thick fog and you can only see as far as the lights of your car enable you and need to move forward a short distance at a time. Here we discuss ways to effectively read and write research papers and aim to help the transition from graduate studies to research.

In the first stage, it is helpful to devote a period of time, about two months, to pure learning of your advisor's research area and the domain of the problem that your advisor suggests. One needs to be interested and become absorbed in the information and like a sponge learn as much as possible. At first your mind might fight the mindset; however, after studying and con-necting to good papers and books, learning the key concepts and observing patterns, the mind starts to pick up the way of thinking and you will see better and pick up information more quickly and deeply. Your mind needs time to get trained on the domain of the problem. After this period of pure learning, understand the problem and think about it as you continue to study for another month or two. During this time it is good to start working on the paper and write the first draft of the introduction. This will help organize in writing what has been achieved in the literature as you study them. It is also

DOI: 10.1201/9781003299073-4

likely to speed up your learning and writing process by seeing that the first two pages of the paper are being completed and a paper is now ready to be added to and developed.

As you study, write ideas that come to your mind on how to approach the problem. The more you read and observe and get into the mindset, the better you will see what needs to be done for your problem and what path needs to be taken to solve it. The papers in well known and high rank journals tend to explain the area well in their first sections making them good choices for model papers to study well.

In contrast to taking a course and concentrating on one textbook, in research one finds many sources available for the same information. When studying one book and not understanding a concept, another book might give more details on that concept and help one understand the material better; however, the mind is likely to become overwhelmed if one switches frequently between different sources and does not allow sufficient time to understand and get into one source at a time. It is best to decide at the beginning which books and papers are mostly related and make them the focus of the study to form a base and to obtain ideas necessary to formulate a plan for the problem. As one becomes stronger in the domain of the area, then he/she can expand to other sources and broaden his/her study on the subject.

Knowledge is like food for the mind. When you are catching up in learning a research area, the mind would be willing to absorb information. Feed your mind the right food and allow it the time to digest. Deep, theoretical papers need more time to digest and cannot be viewed as fast food. Knowing and understanding the domain of the area well will show in your writing and add to its depth. It helps to organize the information and group the papers and material the right way. This helps the mind categorize and keep up with the information. As you read each paper, write a summary on the concepts and ideas and write without details, the problem that is the focus of the paper and the main steps the authors took to solve it.

When you reach your saturation point in learning, it helps to reflect and review what you have studied thus far and settle your thoughts by learning them well. If one learns on the surface and applies the knowledge as he/she learns to get the problem done, he/she would not gain the strong foundation for future research. The amount of time and effort put on writing a research article is often equivalent to that of taking a graduate course. Research involves both learning and solving problems and the goal should be to learn the material for more depth and not just solve the problem. To write a good paper, half of the effort has to be on learning.

Unlike graduate courses, there are no written exams in research and one needs dedication and discipline to learn the material well as he/she conducts research. Giving talks can be treated as oral exams and writing papers as written exams; however, one is required to learn much more than what is said and written on these exams for a solid and strong background. Also instead of taking many different courses with different mindsets, as most are trained in undergraduate and graduate programs, in research all efforts and time are put into one particular problem and the mind has to get centralized to focus on only that problem. To lighten this pressure, it helps to attend mathematics research talks offered by the university or national conferences and learn and be interested in other research topics in your area and field. In addition, it is helpful to attend talks in other departments such as philosophy or history or other sciences like chemistry or engineering to stimulate your mind and keep it interested in learning and thinking about new concepts. Each department at a university often posts on its website information on upcoming talks and lectures. Many research papers in mathematics include applications to other scientific fields to strengthen their results and thus learning and attending talks in various scientific areas would keep your well-rounded interest in science and make the formulation of applications in future papers less difficult. The key is to balance between pure and applied mathematics and work and be interested in both to ensure stronger results and more opportunities in the future.

To make the learning stage in research less time consuming, it helps to organize and group the related papers and books that were found based on style, idea, type of technique used, or by same author and study one group at a time. Also to ease the transition from books to papers, you may ask a printing shop to make a book out of the papers you have chosen to study. Places such as FedEx can make copies and bind the papers in a book form. Having a hard copy of model papers together in one place to focus on is very helpful in becoming accustomed to research papers since it is easier to read them on printed hard copy than on a computer screen.

Most research articles can be categorized into two types of writing styles. One is focused on one particular topic throughout the paper and is self-contained relying mostly on fundamental concepts and theorems in the area of study. The other type does not limit the number of sources used and brings in many different topics and techniques to achieve the result. Typically books, especially textbooks, are written by the first type and they tend to draw the reader in and have an instructional tone to them. The second kind is more research–based and some time needs to be devoted to find the sources they refer to and is more difficult to get into. Both types are important to become accustomed to and observe from.

To study each paper, first get some exposure to it by observing its length and style and determine what it is about based on the title and abstract. Then after a day or two, read it in more depth. This gives the mind the time to adjust to the level of the paper. Read the introduction and main results and write a summary in regular language without calculations or details on what has been done in the paper as a whole and in each section and what are the main ideas before studying the details. The key is to separate ideas from calculations and not get lost in the details.

Zoom in and zoom out of the paper, to focus on both the big picture and the details. To make the reading more active, it helps to write notes and rewrite important passages to study from. For each paper that I plan to write, I devote one notebook for brainstorming and thinking and another notebook for studying and rewriting papers with more steps. Since most research papers are dense, it is useful to rewrite the papers and read as you write. To have a thinking notebook on the side as you study papers and write ideas that you begin to think about is very useful. This notebook can be designed for scratch work and where one does not have to be perfect and can be creative on the ideas and calculations that he/she can think of. It helps to overcome writing blocks and keep the calculations and ideas flowing. I also write the drafts of the calculations in this notebook and use it to write notes to myself on places in the paper that I have doubts on and need to think more about and improve. This notebook helps me to think actively about the problem and plan the content of the paper that I am writing. Having a place where one can think creatively, keep up with ideas and connect the dots the right way is important, especially in mathematics, in which accuracy is the top priority and aiming for perfection can limit one's creativity.

Keep writing ideas as you study and read papers and take notes on how each paper approached a similar problem. At the beginning, the problem seems impossible but as you put blocks of time to only study and think about the problem, you will begin to see patterns in papers and get ideas on how to approach the problem. When studying related papers you are acting as an investigator, trying to find what has been done and approaches taken and then when writing the introduction of the paper, you are acting as a journalist, reporting on your findings in a way that is organized and understandable. Cite the similar problems you find in your paper in the introduction and get ideas from them. Be determined of your own problem and way of proving it. Do not lose track of your own problem nor become discouraged to think that it has already been done. Let finding more papers on the same type of problem or technique give you confidence on its importance. You have to be realistic and responsible in what you state in the paper, but also have to believe in your own intuition and knowledge. It is important to be brave about your creativity and result by presenting the right reasoning.

Formulating the main idea in solving the problem is the first milestone in writing a paper. Once you see a reasonable path to take based on your observations of other papers and your own reasoning, you can get started on the main content of the paper. Write the main ideas and major steps and see the way to break them up into sections and then work on adding and filling up the details in each section. Instead of thinking of a research paper as one major problem to solve, try to break it up into parts and smaller problems to focus during each period of time. These parts can then be formed into sections and made organized. To find an approach to the problem, sometimes thinking about a simplified version helps one see the way to solve it.

For example, finding and observing how the theorem under study has been proven in the literature for other less involved equations and determining methods to overcome the difficulties in applying the same techniques to the more complicated equation under study. Also if you find a technique that was first introduced in a research article, you may click on the citations of that paper on MathSciNet to find other articles that have cited the paper and thus have most likely used the technique and can be used as more examples to observe and study from.

As you work on a problem, new problems will likely rise that have to be taken care of for the results to be correct. Keep writing these issues on the side in the thinking notebook and do not lose confidence or track of the main idea you had in mind. After writing the first draft of the main steps, time may be put to take care of one or two issues at a time. Being positive and battling negativity throughout the research process are necessary. When you face obstacles and the proof does not seem to work, keep thinking of ideas to overcome them. If you really focus your mind on figuring them out, then you will find a way. Taking care of these obstacles make the paper more involved and adds to its value. Studying related papers from higher level journals are likely to bring some light to your problem and their level of difficulty often make your problem seem easier and its solution more attainable. Attending research conferences and listening to talks that are related, are also very effective in giving you ideas. First see the path and the overview structure on how to solve the problem, then keep adding the pieces to the paper and improving your drafts.

Not seeing a path in solving a problem or facing a major issue in the paper, often feels like a dark cloud weighing on one's head. Having determination and dedicating a period of time to get absorbed and actively think through it, would help find a path and overcome the barriers and thus brighten the cloud and reduce its weight. One still has to keep up with the cloud throughout the process and not lose his/her train of thought while attending talks, teaching or studying other papers. Keeping some keywords on your mind can help to remind you of the problem as you tend to other daily activities. It is important to hold on to the problem and get into it and make it part of your mind. To keep thinking about it, finding information and being focused on its steps and writing the pieces. Let observing model papers help you write well. Each paper has to bring some light and understanding to the literature.

Some papers bring new light by introducing a new technique and some strengthen that light by applying the technique to different problems. The papers that offer a new method that can be applied to many problems, open a door to more research and more papers to be written by various authors. For example, the article titled, "Large Deviations for Infinite Dimensional Stochastic Dynamical Systems" by A. Budhiraja, P. Dupuis, and V. Maroulas introduced a new technique to prove a limit theorem called the large deviation principle and since its publication in 2008, over one hundred papers have applied this technique to various problems.

In research, the greater number of citations a paper has, the more it is valued. Authors often take pride in having many citations for their papers. Having numerous citations can prove the paper's usefulness and impact in the community. Sometimes introducing a new concept and building its ground work in a paper generates more papers by encouraging others to develop this idea into an area of study. The articles that have the potential of opening doors and encouraging others to turn them into more articles of different views and settings are more likely to be accepted in higher level journals. For example, the article on large deviations mentioned above, was published in *Annals of Probability*, which is the top journal in probability.

Some papers act as transitions from purely theoretical papers to applied fields and help promote the use of theoretical results in other areas of study. Some through experiments and observations model a physical phenomena like tumor growth in cancer or relate it to an equation that is well known with the right parameters. Some take the model that has already been established and accepted and prove mathematical theorems on it like limit problems by setting a parameter to go to zero or establish the existence and uniqueness of solutions. They apply the theorems to that equation and then describe what the results mean in the context of the physical phenomena. They provide information with mathematical proofs to scientists who work on the phenomenon in labs and can simulate and confirm the results by experiments.

In the process of writing the paper, each statement has to be perfect and needs to be verified by the right reasoning or a reference to another paper or book needs to be made to justify its validity. After accuracy, the concentration should be on organization. A good paper is organized and one can look through it and in a short period of time be able to find the ideas and main results. The paper needs to engage the reader and flow the right way. It is best to work on each piece and section at a time, but at the end, make sure that the pieces fit well together and are written following the same style. It should not be in pieces without the right transitions. Being organized in one's thoughts, shows in his/her writing. In each section, state the ideas and don't let them be lost in the details. One should also use the proper language for mathematics research papers to connect with the reader. At the end of this chapter, a list of commonly used key words is given that can help refresh your mind on the types of words to use in the paper. It is best to balance between equations and words and not have too many equations in one place.

When working on the content of the paper, one should be very careful not to plagiarize and use someone else's work without acknowledging it. Even self-plagiarism, which is using your own statements in a previously published paper, is looked down upon. One needs to learn the information and reword and organize it in his/her own way. The paper should be one's own work and ideas, adhering to the proper language and reasoning in mathematics. Since mathematics relies heavily on precision, it is sometimes difficult to change a definition and it can be given as it appears in well-known books. Some theorems such as the dominated convergence theorem do not need any reference, however, if you are using an idea from another paper or a theorem that you think a regular reader of the article in the field might not be aware of, then a reference needs to be given. For an idea one might say, "inspired by [.], we proceed as follows" or "we use the idea introduced in [.] to obtain,".

As you work on a paper and study related papers more deeply, keep observing what has not been done in the literature and make a note of the problems that seem reasonable to solve in the future. In research one has to devote much time and effort to each article, but also keep the cycle of writing active and have plans on what to work on next. When a paper is complete and to the best stage it can be, then submit it to a journal and put it aside and move to the next paper. Don't let your mind get saturated on old papers and information and waiting to hear back. If the paper gets rejected, then send it to another journal. If you see a purpose in the paper, have observed papers like it that have been published and have put the right amount of time and effort on the results, then don't put it away or devalue your accomplishment. Believe in your reasoning and find another journal that is likely to see its value and appreciate the results.

In research one needs to attain and strengthen the ability to:

- learn many concepts in a short time and to organize the information,
- to stop time and think clearly about a problem,
- to find a good problem, believe in it and be determined to solve it,
- not get drowned in the information and stay on top and focused on the problem,
- find what is needed for the problem from other papers and books and also learn what is needed for more depth and background,
- come up with new problems along the way,
- to have all the information one has studied in mind and connect the dots the right way to find a path to take,
- to have complicated ideas but write them in an organized way for readers to follow and understand the approach,
- to work and focus on one piece at a time and also have all the pieces in mind in the background.

Make each paper you are working on an improvement of the previous paper and your best paper yet. Keep working on drafts and improving the paper and at the end focus on its language and the flow. As in the popular saying, it is about the journey and not the destination. Each paper has great value and requires learning many concepts and making observations in the domain of the area of study. Keep learning and writing and improving the paper. It is not possible to write a research paper over night and it requires time, effort and patience.

Research helps you grow in mathematics and become stronger in understanding and thinking in mathematical problems. Sometimes after grading many exams or preparing lecture notes for the classes that you teach, you seem to lose your train of thought in research. Balance your mind and remind yourself of the problem by again studying the model papers. Help others by teaching and at the same time help strengthen your own abilities in mathematics by conducting research and being involved and active in the mathematical community.

4.1 Common words and Expressions for a Mathematics Research Paper

Establish
establish
achieve
accomplish
examine
study
investigate
analyze
explore
illustrate
demonstrate
determine
discuss
prove
show
derive
obtain
verify
goal, aim, purpose
Desired Result
Principal Result
Main Result

Set
set, let, denote
define, take
introduce

Methods
methods
techniques, tools
strategy, procedure, apply

Importance
well-known
widely used
important, significant, essential
frequently used
classical result
commonly studied
generally
often
usually, typically
by convention
as shown in the literature

Implying
note that, notice that, observe that
we point out that
use the fact that
we obtain
we have
we arrive at
implies
yields

leads to
as a result
as a consequence
due to
giving
it follows that
we can conclude that
we can deduce that
according to
enables us to

Conditions
conditions
assumptions
hypothesis
properties
estimates
calculations
computations
exceptions
required
hold
satisfy
guarantee
imply
fulfill
follow
aid, help to obtain, ensure
allow or permit us to obtain
will enable us
based on
according to,
by
applying the condition
assume
suppose

consider
impose
utilize
include
since
implement
suitable
appropriate
criteria
it is sufficient to show
need to show
it is not difficult to see
it is easy to show
technical details
equivalent to
continues to hold
corresponding to

Between Statements
Therefore,
Thus,
Hence,
Namely,
i.e.
in other words,
in particular,
more precisely
alternatively
moreover,
furthermore,
however,
otherwise,
similarly
analogous to

5

Preparing Documents in Latex

An introduction is given here on how to use the computer program Latex to prepare research papers, dissertations and presentations in mathematics. Latex may be downloaded free of charge from,

www.MikTex.org

One needs an accompanying program in which to type the document. Because of its organized structure, WinEdt is recommended and it can be downloaded from,

www.winEdt.com

by paying a one time registration fee. If you do not wish to use WinEdt, you may download the free of charge software offered by MikTex to type documents. Here we have focused on how to type and compile in WinEdt with examples at the end of the chapter. Note that exactly the same codes may be applied to prepare documents in MikTex.

5.1 The Beginning Code

To start, go to WinEdt and open a new blank file. The first line in the file should be the document class. For example,

\documentclass[12pt,reqno]{amsart}

where 12 is the font size. The document class "reqno", as above, is commonly used for research papers. This document class will place the equation numbers to the right of the equation. For example,

$$ax + b = c \qquad (4.7)$$

Document class "article" can also be used for research papers. For books and dissertations, the document class "book" is used. One may type,

\documentclass[12pt,book, reqno]{amsart}

DOI: 10.1201/9781003299073-5

After the document class, one needs to indicate which packages Latex should use. The following are the common list of packages:

\usepackage{amsmath, amsfonts, amsthm, graphicx, color, amssymb, cite}

You may either give the names of packages in separate \usepackage{..} or group them in one \usepackage as above with commas separating different packages. The above packages offer the typical setup one needs for a mathematics paper. If you need a special feature or code that these packages do not provide, you may add the package to the list. If Latex does not recognize the name, search in your computer for MikTex admin. and in the admin. you can find the package and click on the plus sign on top to include it in your Latex directory. Afterwards, Latex will recognize the package and you may include it in the list of packages and use its features. For example, package "cite" is helpful in organizing references. The package "geometry" is useful when you need to change the margins of the document. For dissertations and books, often you need to follow a specific format and you may do so with this package. For instance,

\usepackage{geometry}[left=2.5cm, right=2.5cm, top=3cm, bottom=3cm]

indicates the length of each margin. For dissertations or other documents that have strict guidelines on margins, it is best to print one or two pages of such documents that have been approved and measure the margins and keep changing the numbers in package geometry to match these lengths. Many times Latex does not give the exact measure of margins as one puts in the code and the numbers have to be played with to obtain the exact margins by printing, measuring and matching.

If one needs the document to be double spaced, he/she may type,

\renewcommand{\baselinestretch}{1.5}

after packages and change the number 1.5 if the space needs to be adjusted. Space number 1.1 is equivalent to single space and is preferred for articles. If no date needs to be given at the beginning of the article under the title, then \date{} may be typed. To make the theorems and lemmas appear in Italic and different from the rest of the text and numbered, the following code is given,

\newtheorem{theorem}{Theorem}

similarly for definition, lemma and proposition. Figure 5.1 gives the exact code needed to be typed at the beginning of Latex file before typing the main document.

FIGURE 5.1
Text to be typed before \begin{document}.

Below we offer an example of how to enter the title and authors' names.

> \title[\emph{CLT for Stochastic Navier-Stokes Equations}]
> {The Central Limit Theorem for Two-Dimensional Stochastic
>   Navier-Stokes Equations}
> \author[D. Jones and A. Webber]{David Jones and Andrea Webber}

The abbreviated title and authors' names are what appear alternatively on top of the pages of the article.

When typing the main body of the document, one has to indicate where an equation, example or proof begins and where it ends. If you begin something then you must end it for Latex to compile. After the code including the documentclass and packages,

$$\begin{document}$$

starts the document and at the end of the file, the code,

$$\end{document}$$

ends it and needs to be the last line in the file. Underneath \begin{document}, type,

$$\maketitle$$

for Latex to include the title of the paper in the PDF. Then the abstract is given for the paper by typing,

\begin{abstract}
your abstract is typed here
\end{abstract}

Afterwards, the Mathematics Subject Classification and Keywords are stated. One may find details on these in Chapter 2. The first section begins with the title,

\section{Introduction}

which produces,

1. INTRODUCTION

in "reqno". If you wish the titles of the sections to be in lowercase you may use,

\section{I \ lowercase{ntroduction}}

If one does not wish to number the sections, he/she may indicate it by a star (by pressing the shift key and number 8 on the keyboard). For example,

\section*{Introduction}

compiles to,

INTRODUCTION

To compile the Latex file, the icon on top of the file with "T" and "PDF" beside a lightening symbol has to be clicked on. The place is indicated by an arrow in Figure 5.2.

FIGURE 5.2
WinEdt Main Page.

As one types the document, he/she may compile to see how the document looks in PDF. The first time the button is clicked on, Latex will ask where to save the document and the name to attach with the file. Afterwards, Latex will start compiling. If you see, "PDFTeXify Compilation Report" with the number of Errors, Warnings and Bad Boxes, then Latex has compiled and the PDF will appear. The PDF is automatically saved in the same folder as the Latex file and can be found there if the PDF does not appear after Latex has compiled. If there is an error, Latex will let you know at which line the error has occurred. You may right click on the horizontal bar next to the text and click on "Show Line Numbers" to find the correct line number. Before compiling again, make sure to close the PDF file if the compiling was successful or click on the red x above the compile report if it was not successful, since Latex can produce one PDF file or report at a time. If the compile button is clicked when the previous time was not successful, an error message will appear indicating that another application is running. Clicking the red x above the compiling report, would end the previous compiling and let one compile again. As mentioned earlier, in most cases errors occur when one does not close a parenthesis or \begin{...} and not end or not end with the exact same command.

In case of a missing parenthesis, braces or bracket, you can place a closed parenthesis, braces or bracket at the end of the equation where the problem is and Latex will highlight the open parenthesis, braces or bracket that has not been closed. If the document is long, then you can start a new document and copy and paste parts of the paper or equation at a time and compile that document to determine where Latex has problem compiling and zoom in on that problem by deleting everything else in the new document. If the error asks for "File Name" then copying all of the document and pasting it in a new blank file and compiling it under a new name usually takes care of it. Sometimes the error is due to Latex not having the right package so the error indicates that the symbol or command is not recognized and one can include the package by going to MikTex admin. as discussed above.

5.2 Typing the Main Content

For the content in a Latex file, one can type in the same way as in a regular Word document with a few extra rules. To start a new line, the previous line has to end with two backslashes: \\. The key above Enter on a keyboard can be used for these backslashes. For example,

> The line ending the paragraph. \\
> The next line.

If one types the text given above without \\, the next line will compile next to the previous line, even if it is typed on the next line on the WinEdt file. For research papers, there should be no line space between the paragraphs and the start of a new paragraph should be indicated by an indent. To start a new paragraph, use \\ at the end of the previous paragraph and right underneath, begin the new line with \indent. If you do not wish for a new paragraph to start and like to take out the beginning space, you may use \noindent at the beginning of the paragraph. If you wish to start a new page, type \pagebreak at the end of the text where the previous page is meant to finish. For example, if you are preparing a document with sections and want each section to start on a new page, then \pagebreak may be applied to start the next section on a new page. Instead of going to the next page, to make the text appear in two or more columns like in a newspaper, one can use multicols. For example, the following code places the text in two columns,

\begin{multicols}{2}
the text is typed here
\end{multicols}

and \usepackage{multicol} needs to be included in the packages. For documents such as Curriculum Vita (CV), it is helpful to use tabular, as follows,

\begin{tabular}{p{6cm}p{4cm}p{5cm}}
Fall 2019-Present & Title 3 & University Three \\
Fall 2017-Spring 2019 & Title 2 & University Two \\
Fall 2015-Spring 2017& Title 1 & University One \\
\end{tabular}

which compiles to,

Fall 2019-Present	Title 3	University Three
Fall 2017-Spring 2019	Title 2	University Two
Fall 2015-Spring 2017	Title 1	University One

where & is placed before the next column's entry and the length of each column is given next to \begin{tabular} by p{"length"}.

If one wishes to make part of the text bold or in Italics, he/she needs to use \bftext{...} or \emph{...}, respectively. For example, \bftext {Principal} and \emph {Theory}, produce **Principal** and *Theory*, respectively. The size of some of the text within the document may be adjusted by the following sizes available from Latex,

\tiny, \scriptsize, \footnotesize, \small, \normalsize, \large, \Large, \LARGE, \huge, \Huge.

For example, {\tiny{*example*}} compiles to example. To change the color of the text, for example when preparing a response to a referee's comments and wanting to highlight the changes, you may use \textcolor{"*color*"}{"your text"}. For example,

\ textcolor{blue}{Theorem 1}

produces the word, Theorem 1 in blue. If you wish a part of the text to appear in WinEdt and not in the PDF file, place % at the beginning of the starting line of the part to make it into a comment.

To make the text appear in the center, to the right or left, one can use,

\begin{"command"}
the text to be aligned
\end{"command"}

where the word "command" is center, flushright or flushleft,respectively. For example,

\begin{center}
The Main Theorem
\end{center}

produces,

The Main Theorem

and

\begin{flushright}
Q.E.D.
\end{flushright}

produces,

Q.E.D.

To give a theorem and make it Italicized in order to differentiate it from the rest of the article, use,

\begin{theorem}
statement of the theorem
\end{theorem}

For example,

\begin{theorem}
Suppose $\{x_n\}_{n\geq1}$ is a cauchy sequence that is bounded above by the constant, $M > 0$. Then as n tends to infinity, the sequence converges.
\end{theorem}

gives,

Theorem 5.1 *Suppose $\{x_n\}_{n\geq1}$ is a cauchy sequence that is bounded above by the constant $M > 0$. Then as n tends to infinity, the sequence converges.*

One may also write a description or a reference next to the theorem such as,

\begin{theorem}[Banach Fixed-Point Theorem]
If $f : X\rightarrow X$ is a strict contraction and X is a nonempty, complete metric space, then f has a unique fixed point.
\end{theorem}

which compiles to,

Theorem 5.2 (Banach Fixed-Point Theorem) *If $f : X \to X$ is a strict contraction and X is a nonempty, complete metric space, then f has a unique fixed point.*

The description might be a reference, for example,

Theorem 4.12 in \cite{5}

where theorem 4.12 in the fifth reference in the article is being referred to. How to cite and give references in an article are given later in this chapter. To be able to refer to a theorem by the number assigned to it by Latex, you may use \label{"name"} next to \begin{theorem}, where "name" is the name you may choose for that theorem. For example, for the above theorem, we can label it using the code,

\begin{theorem}[Banach Fixed-Point Theorem]\label{Banach}

and afterwards can refer to the theorem in the text by Theorem \ref{Banach} to obtain the exact number of the theorem. The same codes may be used for a lemma, proposition, remark, example or definition. For the proof, use,

\begin{proof}
The proof goes here
\end{proof}

which will begin with *Proof* and will place a box at the end of the proof. If the statement of a theorem or lemma is given without a proof then one can manually place a box by,

\begin{flushright}
$\ \Box$
\end{flushright}

In the text, if an equation or a symbol is typed, it should begin and end with a dollar sign: $. For example, $f(t)$ compiles to $f(t)$. Subscript or superscript should also begin and end with dollar signs. For example, n∧th will not compile and n$\wedge\{th\}$ needs to be typed. If an equation is large enough to be given on a separate line then,

\begin{"equation type"}
your equation
\end{"equation type"}

may be applied with "equation type" being "equation", "eqnarray", "flalign" or "align". If the equation can fit on one line within the margins, then equation is appropriate. For example,

\begin{align}
a∧2 +b∧2=c∧2
\end{align}

compiles to

$$a^2 + b^2 = c^2 \tag{5.1}$$

If there are multiple estimates or long equations that must be formed into multiple lines, then "eqnarray", "flalign" or "align" may be used. For instance,

> \begin{eqnarray}
> f(x) & = & (x+2)∧ 2 + 5x + b \\
> & = & x ∧ 2 +4x +4+5x+b \nonumber \\
> & \ leq & x ∧ 2 + 9x +b \nonumber
> \end{eqnarray}

gives,

$$
\begin{aligned}
f(x) &= (x+2)^2 + 5x + b \\
&= x^2 + 4x + 4 + 5x + b \\
&\leq x^2 + 9x + b
\end{aligned}
\tag{5.2}
$$

The form "flalign" starts the equation and the lines after, at the start of the line without any space and can be used to shift the equations to the left to use the maximum space provided by the margins and make the equations more compact. For example, in contrast with the above,

> \begin{flalign}
> & f(x)= (x+2)∧ 2 + 5x + b & \\
> & x ∧ 2 +4x +4+5x+b & \nonumber \\
> & \ leq x∧ 2 + 9x +b & \nonumber
> \end{flalign}

compiles to,

$$f(x) = (x+2)^2 + 5x + b \tag{5.3}$$
$$= x^2 + 4x + 4 + 5x + b$$
$$\leq x^2 + 9x + b$$

The symbol, & indicates where the next line needs to begin. Here in "flalign", & has to be placed at the beginning and end of each line. In "eqnarray", & may also be placed at the beginning by putting two & at the start of the first line such as,

\begin{eqnarray}
&& f(x) = (x+2)∧ 2 + 5x + b \\
& = & x ∧ 2 +4x +4+5x+b \nonumber \\
& \ leq & x ∧ 2 + 9x +b \nonumber
\end{eqnarray}

giving,

$$f(x) = (x + 2)^2 + 5x + b \qquad (5.4)$$
$$= \quad x^2 + 4x + 4 + 5x + b$$
$$\leq \quad x^2 + 9x + b$$

Before submitting the paper or dissertation that you are working on, it is important to check and make sure that all equations are within the margins of the document. This will make the document more appealing to the reader who evaluates the document for acceptance.

Sometimes when there is an uneven balance between the text and equations on a page, Latex puts extra white space at the end of the page. One way to fix this is to type,

\raggedbottom

after packages before begin document at the beginning of the file. Another way is to use "align" instead of "eqnarray" for the equation at the bottom of the page that might be causing the space.

Similar to theorems, to label an equation to be able to refer to it later in the text, \label{"name"} is used, where "name" is what you will use to refer to the equation. That is, to refer to the equation, \eqref{"name"} is typed. For example,

\begin{align} \label{abceq}
ax + b = c
\end{align}
In equation \eqref{abceq} above, we find that $x = (c − b)/a$.

compiles to,

$$ax + b = c \qquad (5.5)$$

In equation (5.5) above, we find that $x = (c − b)/a$.

For multiple equations within one set such as a system, each line may be labeled such as,

\ begin{eqnarray}
z &=& 2x +5y \label{system eq1} \\
z &=& 6x- y \label{system eq2}
\end{eqnarray}

As in the example of "eqnarray" in system (5.4) above, the command \nonumber is used if a particular line in "eqnarray", "flailgn" or "align" need not be numbered in the PDF. If all the lines in a system do not need numbering then a star is placed by using the shift key and number 8, the same as for sections, discussed previously. For example,

> \begin{align*}
> ax + b = c
> \end{align*}

produces,

$$ax + b = c$$

If you need to add some words within the equation, then \text{"the words"} or \mbox{"the words"} may be used. For example,

> \begin{align*}
> x = 2 \text{ if and only if } b = 5
> \end{align*}

gives,

$$x = 2 \text{ if and only if } b = 5$$

To place more space within a line of text or equation, \hspace{2cm} can be typed, where one can also use inches such as \hspace{2in}. For instance,

> \begin{align*}
> x = 2 \hspace{.5cm} \text{if and only if} \hspace{.5cm} b = 5
> \end{align*}

produces,

$$x = 2 \quad \text{if and only if} \quad b = 5$$

For vertical space, \vspace{2cm} may be applied. The equations typed in "equation", "eqnarray", "flalign", "align" will automatically start on a new line and \\ is not needed at the end of the line before them.

In the paper, to number the equations by section number, for example, (5.21) for the twenty first equation in section 5, add the line,

\numberwithin{equation}{section}

at the beginning of the file after the packages before \begin{document}. The same code may be used for theorems and lemmas by replacing equation in above by theorem and lemma, respectively. Without the above code, Latex will number the equations in the order they appear without referring to sections (same for theorems and lemmas). If the document is a dissertation or a book, one may use the same code above with chapter instead of section. To manually label a particular number for an equation, use \tag. For example,

> \begin{align} \ label{abceq} \ tag{2.14}
> ax + b = c
> \end{align}

gives,

$$ax + b = c \tag{2.14}$$

where abceq is what we use later in the text to refer to the equation. The command tag works in "equation" and "flalign" and it does not work in "eqnarray". If one has to center a multiple lined equation and also use tag, then "flalign" accompanied with "hspace" can be used to manually center the equations.

On top of the file in WinEdt, as can be seen in Figure 5.2, there are sets of symbols available. They include all Greek letters and most mathematical symbols one needs such as $\pm$ or ∇. When clicking on a symbol, one still needs to begin and end the symbol by $ signs. After typing a few drafts in Latex, you will begin to pick up and memorize the text codes for most symbols, especially Greek letters. For instance, \beta, \theta and \Omega produce β, θ and Ω, respectively. The text styles \mathbb and \mathcal can also be found that can be applied for letters for spaces. At the end of this chapter, we give codes for some of the frequently used symbols. These codes may either be used in an equation or given in text surrounded by dollar signs. We now offer some details on a few of these symbols. The code, \xrightarrow{"text"} is used when some text needs to be written above the arrow. For example, $x_n \xrightarrow{n \to \infty} \infty$, has the code,

> x_n \xrightarrow{n \ rightarrow \ infinity} \ infinity

To place an arrow sign above a vector, text code \harpoon{...} is the better choice than \vec{...} and for its use, the following line code has to be included at the beginning of the document after the packages and before \begin{document},

> \newcommand{\harpoon}{\overset{\rightharpoonup}}

For absolute value, the vertical bar on keyboard above Enter may be used and to produce the ö in Hölder, a backslash and then double quotation marks beside Enter on keyboard may be applied: H\"older gives Hölder, where no dollar signs are needed. Series are better to give in an equation and not within a line. For example, if we use the same code in our list of symbols but in \begin{align}...\end{align} we obtain,

$$\sum_{k=1}^{\infty} k^2$$

The same holds true for integrals. \frac{...}{...} is used for fractions with the first set of braces giving the numerator and the second set giving the denominator. For epsilon, the symbol \varepsilon is better to use than \epsilon. The code \varepsilon produces ε whereas, \epsilon gives ϵ. Strictly less than or greater than inequalities may be typed directly from the keyboard. A subscript may be placed by clicking on shift key and then minus sign on keyboard. A superscript or power $\wedge$ may be placed by pressing the shift key then number 6 on the keyboard. To place quotation marks, place two ' (the key before number 1 key) at the beginning and for the last two marks click on ' (key next to Enter) twice. For example, "This is a quote".

Placing parenthesis and brackets at the right place and closing one as you open one are very important in Latex. Many times the reason for Latex not compiling is that a parenthesis or a bracket has not been closed. To enlarge the parenthesis to ensure that it fits the expression within more properly, \left(..... \right) may be applied. For example, the line,

(\frac{e $\wedge$ {2}x}{y $\wedge$ 3}) and \left(\frac{e $\wedge$ {2}x}{y $\wedge$ 3}\right)

produces,

$$\left(\frac{e^2x}{y^3}\right) \text{ and } \left(\frac{e^2x}{y^3}\right)$$

If the content within the parenthesis does not fit in one line, you may end the first line by \right. then start the next line with \left. and end the equation with \right). For example,

```
\begin{eqnarray*}
f(t) & = & 5t + 8t \wedge 2 + \exp \left((t \wedge 3 - 1) \wedge 2 + \cos(4t)
+sin(9t + 10) + \tan(6t \wedge 3 + 9t) \ right. \ \
&& \ left. + 10t + \csc(9t - 3) - 4t\right) \\
& \ leq& 7y + 105
\end{eqnarray*}
```

compiles to,

$$f(t) = 5t + 8t^2 + \exp\left((t^3 - 1)^2 + \cos(4t) + \sin(9t + 10) + \tan(6t^3 + 9t)\right.$$
$$\left. + 10t + \csc(9t - 3) - 4t\right)$$
$$\leq 7y + 105$$

The same codes may be applied for brackets, absolute values, braces, and norms. For braces to appear, use a backslash before it: \{...\}. For example, \exp\{x\} gives, exp{x}. In "eqnarray", the first line either starts with && as in (5.4) or & is placed before and after = or inequality sign as in above. If the next line is a continuation of the first line, such as in above, then && needs to be typed at the beginning of the line to place some space and if the next line needs to be aligned with the first line's equality or inequality sign such as $\leq$ in above, then & is needed before and after the new line's inequality.

For a piecewise function, you can use array as in the following,

> \begin{align*}
> f(x)= \left \{\begin{array}{11}
> 5x+4 & \text{if } x \ geq 1, \\
> 6x-2 & \text{otherwise}
> \end{array} \right.
> \end{align*}

which compiles to,

$$f(x) = \begin{cases} 5x+4 & \text{if } x \geq 1, \\ 6x-2 & \text{otherwise} \end{cases}$$

Beside the \begin{array}, the number of ls (small letter L) indicate how many columns one needs, which for a piecewise function is two. For more examples of equations discussed here see Figures 5.3 and 5.4 below.

```
Greek Symbols International Typeface Functions ( || ) — | + / : — —> — < > = ... AMS ... AMS < > = AMS NOT < > =
Π Π ∫ ≠ ∩ U  N B B   xᵏ  à  â  ā  á  à   √abc  a̅b̅c̅  ā̅b̅c̅  a̅b̅c̅  a̅b̅c̅  (ᵐ)
V ∧ ⊙ ⊗ ⊕ ⊎ C T ℑ   xₖ  ā  ā  ā  ā  ā   √abc  abc  abc  abc  abc  abc̲
bons.tex
```

```
Piecewise:
\begin{align*}
\left\{\begin{array}{11}
u_{tt} - \Delta u = f & \text{in }\mathbb{R}^{n} \times (0,\infty),\\
u = 0, u_{t} = 0 & \text{on } \mathbb{R}^{n} \times $\{t=0\}$.
\end{array}\right.
\end{align*}
```

```
eqnarray:
\begin{eqnarray*}
\int_{0}^{\infty} \int_{-\infty}^{\infty} w \phi dxdt &=& \int_{0}^{T} b(x)dx +
\int_{0}^{\infty}\int_{-\infty}^{\infty}w(b_{\varepsilon}-b)v_{x}^{\varepsilon}dxdt\\
&=& \int_{\tau}^{T} \int_{-\infty}^{\infty} w(b_{\varepsilon}-b)v_{x}^{\varepsilon}dxdt\\
&& + \int_{0}^{\tau} \int_{-\infty}^{\infty} w(b_{\varepsilon}-b)v_{x}^{\varepsilon}dxdt\\
&=:& I_{\tau}^{\varepsilon}+J_{\tau}^{\varepsilon}
\end{eqnarray*}
```

```
flalign:
\begin{flalign}
&|u(x,t)-u(x,\hat{t})| \leq \left|\int_{t}^{T} r(x(s),\alpha(s))ds + g(x(T))
- \int_{\hat{t}}^{T} r(\bar{x}(s),\alpha(s))ds + \int_{t}^{T} g(\bar{x}(T)) dx\right. & \nonumber \\
&\left. - \int_{\hat{t}}^{T} h(\bar{x}(s),\alpha(s))ds + \int_{t}^{T} w(\bar{x}(T)) dx \right|& \nonumber \\
&= \int_{T+t-\hat{t}}^{T} y(x(s), \alpha(s))ds + g(x(T)) - g(x(T + t -\hat{t}) &\nonumber \\
&\leq C|t-\hat{t}| + \varepsilon&
\end{flalign}
```

FIGURE 5.3
Examples of equations.

Piecewise:

$$\begin{cases} u_{tt} - \Delta u = f & \text{in } \mathbb{R}^n \times (0, \infty), \\ u = 0, u_t = 0 & \text{on } \mathbb{R}^n \times \{t=0\}. \end{cases}$$

eqnarray:

$$\begin{aligned} \int_0^\infty \int_{-\infty}^\infty w\phi dxdt &= \int_0^T b(x)dx + \int_0^\infty \int_{-\infty}^\infty w(b_\varepsilon - b)v_x^\varepsilon dxdt \\ &= \int_\tau^T \int_{-\infty}^\infty w(b_\varepsilon - b)v_x^\varepsilon dxdt \\ &\quad + \int_0^\tau \int_{-\infty}^\infty w(b_\varepsilon - b)v_x^\varepsilon dxdt \\ &=: I_\tau^\varepsilon + J_\tau^\varepsilon \end{aligned}$$

flalign:

$$\begin{aligned} |u(x,t) - u(x,\hat{t})| &\leq \left| \int_t^T r(x(s), \alpha(s))ds + g(x(T)) - \int_{\hat{t}}^T r(\bar{x}(s), \alpha(s))ds + \int_t^T g(\bar{x}(T))dx \right. \\ &\quad \left. - \int_{\hat{t}}^T h(\bar{x}(s), \alpha(s))ds + \int_t^T w(\bar{x}(T))dx \right| \\ &= \int_{T+t-\hat{t}}^T y(x(s), \alpha(s))ds + g(x(T)) - g(x(T + t - \hat{t}) \\ &\leq C|t - \hat{t}| + \varepsilon \end{aligned} \tag{1}$$

FIGURE 5.4
PDF of Equations in Figure 5.3.

To include a matrix, either "bmatrix" or "array" may be used. The command "bmatrix" uses brackets to enclose the matrix and "array" uses parenthesis. For example,

```
\begin{equation*}
A = \begin{bmatrix}
2 & r & 4 \\
6 & 0 & 2 \\
7 & 3 & x+y \\
\end{bmatrix} \ hspace{.5cm}
B = \left(
\begin{array}{ccc}
4 & w & t \\
3 & 7 & 8 \\
-1 & 5 & 10 \\
\end{array}
\right) \ hspace{.5cm}
C = \kbordermatrix{
& \ text{P} & \text{NP} \ \
\text {P} & H_{11} & H_{12} \ \
\text {NP} & 0 & 0 }
\end{equation*}
```

gives,

$$A = \begin{bmatrix} 2 & r & 4 \\ 6 & 0 & 2 \\ 7 & 3 & x+y \end{bmatrix} \qquad B = \begin{pmatrix} 4 & w & t \\ 3 & 7 & 8 \\ -1 & 5 & 10 \end{pmatrix} \qquad C = \begin{array}{c} \\ P \\ NP \end{array} \begin{bmatrix} P & NP \\ H_{11} & H_{12} \\ 0 & 0 \end{bmatrix}$$

Instead of placing a "bmatrix" within an equation command, one can also place it between, \[....\]. Another way to use "array" is to click on Insert on top of the file where the symbols are located and click on "Tables", then "Matrix" and indicate the number of rows and columns. This will automatically give the above code. Either c or l (small letter L) can be typed beside "array" to give the needed number of columns. To insert labels for each row or column as in matrix C above, "kbordermatrix" is useful, which requires the following package to be included with the rest of the packages at the beginning of the file before \begin{document},

\usepackage{kbordermatrix}

5.3 Inserting Pictures

To be able to place a picture or figure within the text, include package "graphicx" with other packages at the beginning of the document and then type "includegraphics" in the file as follows,

\includegraphics[width=.8 \ linewidth]{"name"}

where "name" is the name of the picture file. Examples are given in Figures 5.16–5.19. For latex to find the picture or figure, the file has to be saved in the same folder as the latex file used to type the document. In addition, the picture or figure has to be saved with the exact same name as given in "name" in the code above, otherwise, it will not compile. The size of the figure may be adjusted with the number put for width above. For better size adjustment, one also has the option to use,

\includegraphics[width = 17cm, height = 18cm]{"name"}

where, the measure of the width and height may be altered. If the picture needs to be rotated, then the angle of rotation may be given by including angle=... such as,

\includegraphics[width = 17cm, height = 18cm, angle=180]{"name"}

Also one can use text alignment such as center, flushright or flushleft to place the figure in the right position. For example,

```
\begin{center}
\includegraphics[width=.8\linewidth]{"name"}
{"the caption"}
\end{center}
```

will center the picture and place its caption underneath it. Another way to place a caption is to enter \["your caption"\] below includegraphics instead of \caption as in Figures 5.18 and 5.19. To place two or more pictures placed side by side, "minipage" is useful. As in Figures 5.18 and 5.19, for two pictures presented side by side, the following code may be applied,

```
\begin{minipage}[b]{.4 \ textwidth}
\includegraphics[scale=.3]{"name" of the first picture}
\scriptsize{textbf{caption for the first picture}
\end{minipage}
\begin{minipage}[b]{.4 \ textwidth}
\includegraphics[scale=.4]{"name" of the second picture}
\scriptsize{textbf{caption for the second picture}
\end{minipage}
```

As for inserting a table, the following website has many helpful codes and different types of tables that may be formed,

https:\\www.overleaf.com\learn\latex\tables

5.4 References

For references, include "cite" in the packages at the beginning and type,

```
\begin{thebibliography}{18}
your references go here
\end{thebibliography}
```

where 18 is the total number of references. The references are typically placed at the end before \end{document}. Sometimes they are placed before the Appendix in the paper. Each reference begins with,

\bibitem{"name that is used to refer to it"}

then the reference is given. For example,

\begin{thebibliography}{2}

\bibitem{Klebaner}
F. Klebaner (2005). \emph{Introduction to Stochastic Calculus with applications} (2nd ed.). London, England: Imperial College Press (ICP).

\bibitem{Glasserman}
P. Glasserman, W. Kang, P. Shahabuddin (2007). Large deviations in multi-factor portfolio credit risk. \emph{Math. Finance.} vol. 17, no. 3, 345–379.

\end{thebibliography}

To refer to a reference in the text, use \cite{"name" given in bibitem}. For example, \cite{Glasserman} gives [2]. It is convenient to use the first author's name for bibitem. In research articles, the number of reference is given to cite the paper as [.]. In books and dissertations, often the APA style is applied, in which the authors' names and the year of publication are given. For instance, to cite the second reference above, one writes, Glasserman, Kang and Shahabuddin (2007). In APA style, if a paper has one or two authors, then each author's last name is given. If a paper has three to five authors, each author's last name is typed for the first time the paper is referred to and afterwards if one needs to refer to the paper again, the first author's name is given, then et al. is typed. For example,

"For standard introductory textbooks on differential equations, we recommend Boyce and DiPrima (2009) and Polking, Boggess, and Arnold (2018) and for linear algebra we refer the reader to Anton and Rorres (2005) and Larson (2013). We note that in Polking et al. (2018), section 5.3 concentrates on the topic of this paper and may be used as a source for background."

To write the references in APA style, one first gives the last name of each author followed by his/her first initial as shown below,

\begin{thebibliography}{2}
\bibitem{Klebaner}
Klebaner, F. (2005). \emph{Introduction to Stochastic Calculus with applications} (2nd ed.). London, England: Imperial College Press (ICP).

\bibitem{Glasserman}
Glasserman, P., Kang, W., & Shahabuddin, P. (2007). Large deviations in multifactor portfolio credit risk. \emph{Mathematics Finance} \emph{17}(3), 345–379.
\end{thebibliography}

Here, only the first letter of the first word in the title of books is capitalized as in title of papers. In addition, the volume number is in Italic and & is placed before the last author's name if there are more than two authors. For more examples on references see Figures 5.5 and 5.6.

```
ath: Greek Symbols  International  Typeface  Functions  { Ⅱ }  → • / → ... → ... < > = ... AMS ... AMS < > = AMS NOT < > =
) ó ö õ õ õ ò ó  ç ¿  œ æ å ø ł ı  ä ë ï õ ü ÿ ß  à á â è é ê ì í î ò ó ô ù ú û
$ ó õ õ ø ø ø ø ø ø  Ç ¡  Œ Æ Å Ø Ł Ɛ  Ä Ë Ï Õ Ü Ÿ ß  À Á Â È É Ê Ì Í Î Ò Ó Ô Ù Ú Û
} references.tex

⊟ \begin{thebibliography}{5}

  \bibitem{Ahlfors}
  L. Ahlfors (1973). \emph{Conformal Invariants: Topics in Geometric Function Theory}. McGraw-Hill, New York.

  \bibitem{Alon}
  N. Alon and I. Benjamini, and A. Stacey (2004). Percolation on finite graphs
  and isoperimetric inequalities. \emph{Ann. Probab.} vol. 32, 1727-1745.

  \bibitem{Fleming}
  W. Fleming, H. Soner (2006). \emph{Controlled Markov Processes and Viscosity Solutions: Second Edition}. Springer, New York.

  \bibitem{Luczak}
  T. $\L$uczak, B. Pittel, and J. Wierman (1994). The structure of random graph near the point
  of the phase transition. \emph{Trans. Amer. Math. Soc.} vol. 341, 721-748.

  \bibitem{Schramm}
  O. Schramm and S. Sheffield. Contour lines of the two-dimensional
  discrete Gaussian free field. Available at arXiv:math.PR/0605337.

  \end{thebibliography}

  \end{document}

nsole - PDFTeXify ... (Exit Code=0)
PDFTeXify Compilation Report (Pages: 1)
Errors: 0   Warnings: 2   Bad Boxes: 0
```

FIGURE 5.5
Examples of References.

References

[1] L. Ahlfors (1973). *Conformal Invariants: Topics in Geometric Function Theory.* McGraw-Hill, New York.

[2] N. Alon and I. Benjamini, and A. Stacey (2004). Percolation on finite graphs and isoperimetric inequalities. *Ann. Probab.* vol. 32, 1727-1745.

[3] W. Fleming, H. Soner (2006). *Controlled Markov Processes and Viscosity Solutions: Second Edition.* Springer, New York.

[4] T. Luczak, B. Pittel, and J. Wierman (1994). The structure of random graph near the point of the phase transition. *Trans. Amer. Math. Soc.* vol. 341, 721-748.

[5] O. Schramm and S. Sheffield. Contour lines of the two-dimensional discrete Gaussian free field. Available at arXiv:math.PR/0605337.

FIGURE 5.6
PDF of References in Figure 5.5.

5.5 Dissertations

As for preparing a dissertation, since dissertations often have multiple chapters with many pages in each, then one can use a separate document file for each chapter and then apply the "include" command. See Figure 5.7 for an example on the beginning code of a dissertation and Figure 5.8 for the code to include the content of different files. Each chapter file should only contain the main content and not the beginning code as in the usual latex document and it should not end with end document. For example, \chapter{"Title of chapter two"} \ {chap2} may be the first line in file of chapter two, where chap2 is the name given in \include in the main dissertation file to refer to chapter two and the file should be saved with the exact same name, chap2, in the folder where dissertation file is saved. Since latex cannot compile without the beginning code, then each chapter may be saved without compiling and the main file will compile all chapters at once and give errors if any. The best way is to work on each chapter separately as independent documents with the beginning and ending codes and when they are completed and have been compiled without errors, the "include" command in a new file may be used as above. Another way is to manually copy and paste the content of each chapter in the right order in one file with the right beginning and ending codes and compile that file. To exclude page numbers for the whole document, type \pagestyle{empty} before \begin{document}.

```
\begin{document}

\frontmatter
\thispagestyle{empty}

\vspace*{1cm}
\begin{center}
\huge{\textbf{Title of the Dissertation}}
\end{center}
\vspace{7cm}
\begin{center}
\Large{A Dissertation Presented for the\\
Doctor of Philosophy\\
Degree\\
Name of the Affiliated University}
\end{center}
\vspace{2cm}
\begin{center}
\Large{Author's Name\\
Date (Month Year)}
\end{center}
```

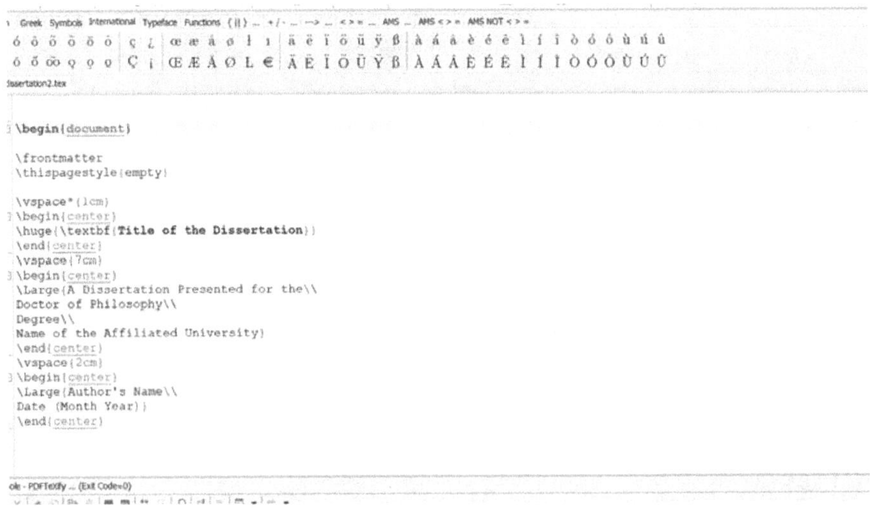

FIGURE 5.7
Code for beginning of dissertation.

If only specific pages do not need numbering, for example the title page and the first few pages in a dissertation, then \thispagestyle{empty} can be typed before the start of those pages. See Figures 5.7–5.11 for the main code of dissertation and its PDF.

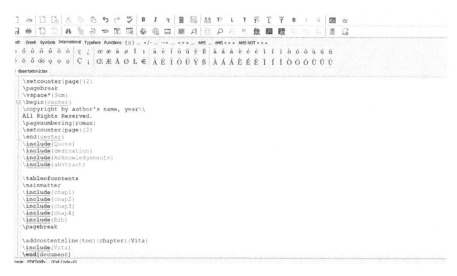

FIGURE 5.8
Code for the body of dissertation.

FIGURE 5.9
PDF of first page of dissertation.

©by author's name, year

All Rights Reserved.

FIGURE 5.10
PDF of second page of dissertation.

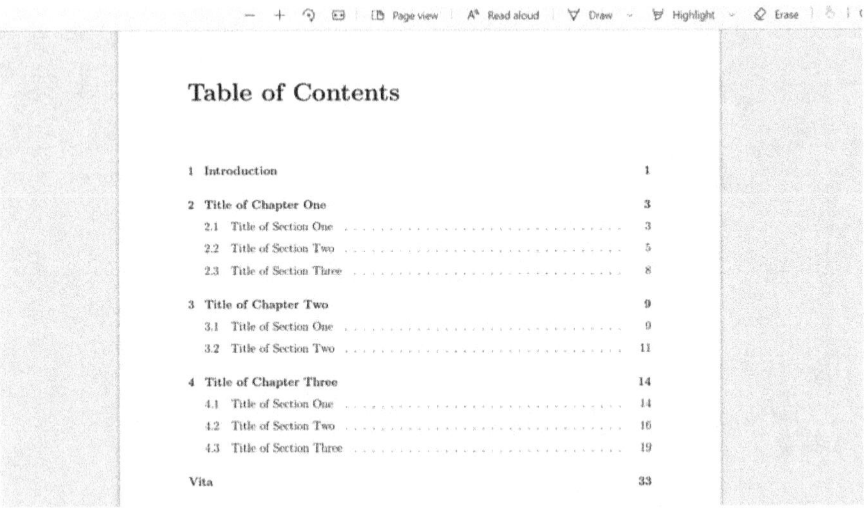

Table of Contents

1 Introduction 1

2 Title of Chapter One 3
2.1 Title of Section One 3
2.2 Title of Section Two 5
2.3 Title of Section Three 8

3 Title of Chapter Two 9
3.1 Title of Section One 9
3.2 Title of Section Two 11

4 Title of Chapter Three 14
4.1 Title of Section One 14
4.2 Title of Section Two 16
4.3 Title of Section Three 19

Vita 33

FIGURE 5.11
PDF of table of contents in dissertation.

5.6 Beamer for Presentations

We now give a description on how to use the document class Beamer to prepare slide presentations for research talks. At the beginning of the file, type the following documentclass:

```
\documentclass[xcolor=dvipsnames]{beamer}
```

For every file there should be only one document class, but there can be multiple packages underneath the document class. The package "graphicx" needs to be included in the packages to be able to insert pictures in the presentations. See Figure 5.12 for the beginning code for Beamer. There are many styles available to use for Beamer presentations. Most Beamer styles have color bars on top of each slide with the title of the slide being in a different color. The code for the theme, Copenhagen used in Figures 5.12–5.19, does not have a solid bar and I find it more proper for research presentations. Other themes may be found online, for example at

https:\\deic−web.uab.cat\~iblanes\beamer_gallery\index_by_theme.html

As in a regular Latex file, after the documentclass and packages, one types \begin{document}. It is helpful to type \end{document} at the end of the document after placing enough space between to type your material. The extra space will not show in the pdf after compiling. After begin document, each slide is given by,

\begin{frame}
the content of the slide
\end{frame}

Changing the font of each slide to \footnotesize or \scriptsize will likely to enhance the appearance of the text and equations within the slide. For example, each slide may be typed as,

\begin{frame}{title of the slide}
\footnotesize{
the content of the slide}
\end{frame}

To change the font of some text within the document the same font sizes mentioned earlier for a regular Latex document may be applied.

To organize the content of the slide into bullet points, the code "itemize" or "description" may be used. See Figures 5.14 and 5.15 for examples. "Itemize" gives the regular format of bullet points. Within a slide, "itemize" may be typed as follows,

\begin{itemize}
\item "type the first item"
\item "type the second item"
\end{itemize}

where each item stands for one bullet point. If one wishes to insert a bullet point him/herself then $\Bullet$ may be typed at the beginning of the content that needs a bullet point. Another code is "description" given by,

\begin{description}
\item[title for item one] item one
\item[title for item two] item two
\item[title for item three] item three
\end{description}

As illustrated in Figures 5.14 and 5.15, in "description", the title of each bullet point is given in Italics and in color blue. In addition, to number items in a list, "enumerate" may be used, which in a regular document will simply number the items on the list instead of placing the numbers in a colored circles as in Figure 5.15. The above codes may be applied in a regular document for a paper to give bullets or enumerate.

Within the slides, equations may be typed using the same codes as in a regular Latex document. The star by the equation type needs to be given such as \begin{align*} to avoid labeling equations on slides. Also pictures and figures may be included by "includegraphics" following the same instructions given in Section 5.3. As mentioned in that section, the picture has to be saved in the same folder as the Beamer latex file. To state a theorem, \begin{theorem}... \end{theorem} may be used. The same code may be applied for definitions, lemmas and propositions. For a research talk, for each theorem, lemma or proposition one states the authors' last names and the year of publication as in the example below,

\begin{theorem}[Abel-J.-Weber '08]
The statement of the theorem
\end{theorem}

where the speaker's last name is for example, Jones and other authors' last names are Abel and Weber and the year of publication or completion of the result is 2008. Thus, only the initial of the speaker's last name is given. In most Beamer themes, theorems, definitions, lemmas and propositions are given in a colored box separate from the rest of the content on the slide. To compile the Beamer presentation, the same icon as in a regular document may be clicked on (see Figure 5.2).

We believe that the information given in this chapter is sufficient to get one started on his/her first draft in latex. For more codes and help on different features available in latex, we recommend the website,

https://www.overleaf.com/learn

Code:	Compiles to:
x \ leq y	$x \leq y$
x \geq y	$x \geq y$
f: X \rightarrow Y	$f : X \rightarrow Y$
g: X \xrightarrow{D} Z	$g : X \xrightarrow{D} Z$
X \overset{d}{=} Y	$X \overset{d}{=} Y$
x \mapsto y	$x \mapsto y$
\nabla u	∇u
x \in \ mathbb{R}	$x \in \mathbb{R}$
(A \cap B) \cup C	$(A \cap B) \cup C$
	(Continued)

Code:	Compiles to:
A \subset B	$A \subset B$
\mathcal{C} ([0,T]; \mathbb{R})	$\mathcal{C}([0, T]; \mathbb{R})$
\mathbb{N}	$\mathbb{N}$
\mathcal{F} and \mathcal{G}	$\mathcal{F}$ and $\mathcal{G}$
L$ \ acute{e}vy, It \ hat{o}$, H\"older	Lévy, Itô, Hölder
\frac{x∧2}{y}	$\frac{x^2}{y}$
{x_n}_{n \ geq 1}	${x_n}_{n \geq 1}$
\sum_{k=1} ∧ {\infty} k ∧ 2	$\sum_{k=1}^{\infty} k^2$
\int_{-\infty}∧{\infty} x∧2 dx + \int 2dx	$\int_{-\infty}^{\infty} x^2 dx + \int 2dx$
\lim_{n \ rightarrow \ infty} \ frac{2}{n − 4}	$\lim_{n \to \infty} \frac{2}{n-4}$
\log{3x} and \exp{(2x)}	$\log 3x$ and $\exp (2x)$
\frac{\partial f}{\partial x}	$\frac{\partial f}{\partial x}$
|x| and |y|	$|x|$ and $\|y\|$
< x > or \langle x \ rangle	$< x >$ or $\langle x \rangle$
\sqrt{4x + 8} and \sqrt[3]{4x + 8}	$\sqrt{4x + 8}$ and $\sqrt[3]{4x + 8}$
\vec{u}, \harpoon{x} and \overrightarrow{z}	$\vec{u}, \overrightarrow{x}$ and $\overrightarrow{z}$
\bar{xy} and \overline{xy}	$\bar{xy}$ and $\overline{xy}$
\tilde{W} and \widetilde{W}	$\tilde{W}$ and $\widetilde{W}$
f_x (x, t)\biggr\rvert_{t = 0}	$f_x(x,t)\Big\rvert_{t=0}$

FIGURE 5.12
Text code for beginning and title slide for Beamer.

FIGURE 5.13
PDF of text code in Figure 5.12.

```
}\begin{frame}
\footnotesize{
  \begin{center}
  \textbf{\large{\textcolor{Violet}{Title of the slide}}}
\end{center}
}\begin{itemize}
\item the first bullet point
\item the second bullet point
\item the third bullet point
\end{itemize}
}\begin{center}
\noindent\rule{8cm}{0.4pt}
\end{center}
}\begin{description}
  \item[The first item] The content of the first item
  \item[The second item] The content of the second item
  \item[The third item] The content of the third item
\end{description}
}\begin{enumerate}
\item first item on the list
\item second item on the list
\item third item on the list
\end{enumerate}}
\end{frame}
```

FIGURE 5.14
Examples of lists.

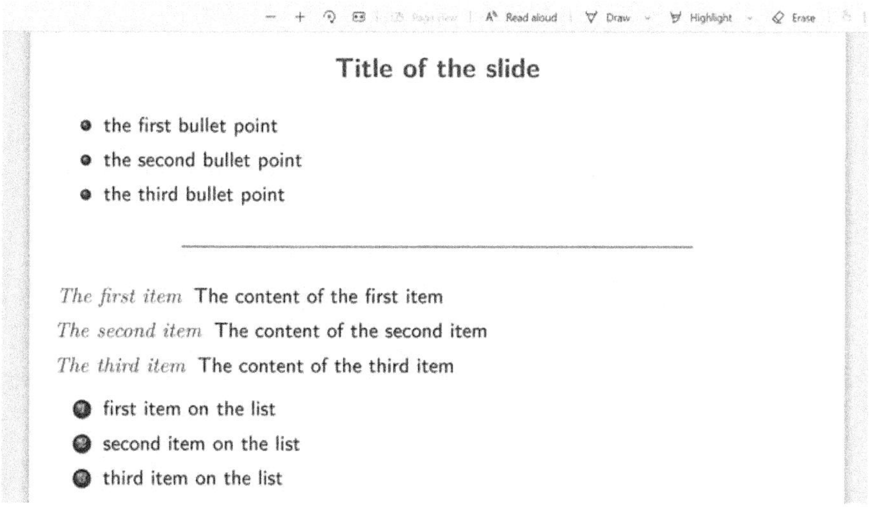

FIGURE 5.15
PDF of lists in Figure 5.14.

```
\begin{frame}
\footnotesize{
  \begin{center}
  \textbf{\large{\textcolor{Violet}{Exit Problem for Some Fluid Models}}}
\end{center}
\textbf{Fluid Models:}
\begin{itemize}
\item Navier-Stokes equation:
\begin{equation*}
\partial_{t}u - \nu \Delta u + u\nabla u + \nabla p = f
\end{equation*}
\item Boussinesq equation:
\begin{eqnarray*}
\partial_{t}u + u\nabla u - \nu \Delta u + \nabla p &=& \theta e_{2} + f\\
\partial_{t} \theta + u \nabla \theta -u^{2} - \kappa \Delta \theta &=& g
\end{eqnarray*}
\end{itemize}}
\begin{center}
\includegraphics[scale=.1]{Exit}
\end{center}
\end{frame}
```

FIGURE 5.16
Text code for a slide with picture on Beamer.

Exit Problem for Some Fluid Models

Fluid Models:

- Navier-Stokes equation:

$$\partial_t u - \nu \Delta u + u \nabla u + \nabla p = f$$

- Boussinesq equation:

$$\partial_t u + u \nabla u - \nu \Delta u + \nabla p = \theta e_2 + f$$
$$\partial_t \theta + u \nabla \theta - u^2 - \kappa \Delta \theta = g$$

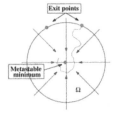

Parisa Fatheddin *Joint Work with ....*   Asymptotic Limits for some Stochastic PDEs

FIGURE 5.17
PDF of code in Figure 5.16.

```
pictures.tex
\begin{frame}
\begin{center}
\textcolor{blue}{\textbf{Applications of Lasers}}
\end{center}
\begin{figure}
    \centering
    \begin{minipage}[b]{0.4\textwidth}
        \includegraphics[width=.5 \textwidth]{telescopes}
        \[Telescopes\]
    \end{minipage} \hfill
    \begin{minipage}[b]{0.4\textwidth}
        \includegraphics[width=\textwidth]{LaserDefense}
        \[Defense\]
    \end{minipage} \hfill
    \begin{minipage}[b]{0.4\textwidth}
        \includegraphics[width=\textwidth]{LaserMedicine}
        \[Medicine\]
    \end{minipage} \hfill
    \begin{minipage}[b]{0.4\textwidth}
        \includegraphics[width=\textwidth]{Metal}
        \[Welding\hspace{.05cm} of\hspace{.05cm} Metal\]
    \end{minipage}
\end{figure}
\end{frame}
```
Console - PDFTeXify ... (Exit Code=0)
PDFTeXify Compilation Report (Pages: 1)

FIGURE 5.18
Code for slide with multiple pictures.

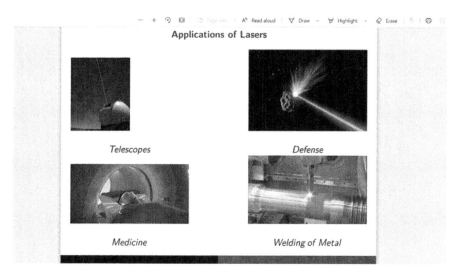

FIGURE 5.19
PDF of code in Figure 5.18.

Part II

Modern Applications in Industry

6

Schrödinger Equation in Laser Technology

Laser technology has taken an important role in medicine, making many medical procedures possible and is widely used in various technological devices such as in telescopes and in regular devices for instance, shopping scanners to read bar codes or in CD and DVD players. The field of image processing employs lasers to produce images with high quality and in the field of remote sensing, lasers help to determine information about the atmosphere, such as the condensation of water in the clouds. Lasers are also used to precisely cut and draw delicate patterns on glass and metals. In order to add to their precision, researchers often study lasers by setting up a medium in which a laser is being shot from a transmitter and investigate how the atmosphere and different obstacles in the air affect the laser beam as it travels to reach the receiver. Different apparatus such as lens, mirrors and prisms are used in the medium to alter the behavior of the beam to obtain the desired form at the receiver. Convex lens, for example in magnifying glasses, are used to focus the rays of laser beam and intensify its energy as shown in Figure 6.1 below. On the other hand, Concave lens, with the shape illustrated in Figure 6.1, scatter the rays and defocus the light.

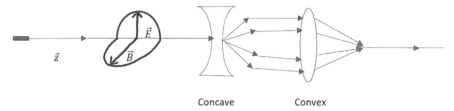

Concave   Convex

FIGURE 6.1
Laser Propagation.

Focusing the laser beam and intensifying its energy to a smaller point is essential in surgical procedures, for example, when some tumor cells need to be destroyed. Self-focusing of the laser beam has been studied in the literature and it describes the situation, in which the rays in the beam converge automatically when traveling in the medium because of the variations in temperature in the medium or the presence of obstacles in the atmosphere.

DOI: 10.1201/9781003299073-6

Lens, mirrors and prisms, as shown in Figure 6.2 below, are the commonly used tools in modifying the laser beam to focus, defocus, scatter, spread or cause diffraction (bending of the light rays). There are two types of scattering: backscattering (illustrated for the lens in Figure 6.2) and forward scattering (shown for prism in Figure 6.2), in both of which notable research has been conducted in the literature on optics. Note that in Figure 6.2, the light rays hit the convex lens at an angle and thus scatter, instead of focus the rays.

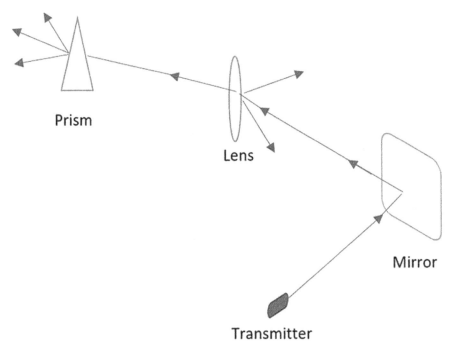

FIGURE 6.2
Apparatus used for Lasers.

Researchers typically use a mirror to reflect the light, which sends the rays at the same angle symmetric to the angle hitting the mirror and does not cause the laser beam to lose any energy. The intensity of the laser beam, referred to as its irradiance, can be reduced by absorbtion, when the atmosphere absorbs heat from the laser. The precise definition of irradiance is the time average of the power of the laser beam per unit area. The changes in irradiance of the beam due to turbulence in the atmosphere are called scintillation. Dispersion occurs when rays of the beam travel at different speeds. Furthermore, prisms are typically used to alter the direction of the laser beam and in the case of image processing, change the orientation of the image.

If the laser is being experimented in a vacuum, then the laser beam is not altered as it travels; however, the primarily interest is in examining the laser beam as it travels in random media, where air, dust, and temperature changes in the atmosphere affect the laser beam. This type of media, that is not homogeneous, is often referred to as turbulent media and the specific places where there is turbulence are referred to as eddies. In image processing, the eddies can cause the image at the receiver to become blurry, referred to as aberration. Based on their size, eddies are grouped in the category of innerscale of turbulence denoted as ℓ_0, outerscale of turbulence denoted as L_0, or in inertial subrange if their size is in between innerscale and outerscale. For example, on the earth's surface, the innerscale contains eddies of size ranging from 1mm. to 10mm. and the eddies that are hundred meters long belong to outerscale.

Reynolds number is typically used to determine if a medium is turbulent. In the study of fluids, the Reynolds number is given by Re= $\frac{VL}{v}$, where V is the velocity of the fluid, L is the reference length (for example the diameter of the pipe through which the fluid is flowing), and v represents the viscosity of the fluid, describing how thick the fluid is. For each type of fluid, if the Reynolds number exceeds the known critical Reynolds number, then the fluid is considered to be turbulent.

Turbulence in a medium is mainly developed by fluctuations in temperature and pressure. These fluctuations are characterized by the medium's refractive index, denoted by n. The index of refraction is determined by $n = \frac{c}{v}$, where $c = 3.0 \times 10^8$ m/s is the speed of light in a vacuum and v is the velocity of light wave in the medium. As a random variable at point $x \in \mathbb{R}$ and time t, it is written in the form,

$$n(x, t) = 1 + n_1(x, t), \tag{6.1}$$

where $n_1(x, t)$ satisfies $\mathbb{E}(n_1(x, t)) = 0$ and is the random deviation of the refractive index, $n(x, t)$, from its mean, $n_0 = \mathbb{E}(n(x, t)) \approx 1$. When there is no dependence on time, the medium is called a frozen turbulent medium, where eddies are assumed to stay in one place throughout time.

The index of refraction is an important characteristic of a medium, since changes in the refractive index from one medium to another are the main causes of diffraction of light. This is depicted in Figure 6.3, where the horizontal line is called interface, separating the two mediums of different refractive indices. The ray moving toward and the ray moving away from the interface are called the incident and refracted rays, respectively.

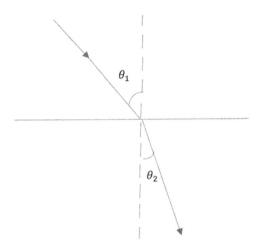

FIGURE 6.3
Diffraction.

In Figure 6.3, if n_1 is the refractive index of the top medium and n_2 is the refractive index of the bottom medium, then by Snell's law the relation between the angles θ_1 and θ_2 is given by,

$$n_1 \ sin\theta_1 = n_2 \ sin\theta_2.$$

It is known that if $n_1 < n_2$, then the angle of refraction, θ_2 is less than the incident angle, θ_1 and the refracted ray is closer to the dotted vertical line. On the other hand, if $n_1 > n_2$, then the refracted ray bends away from the normal line. For example, Figure 6.3 may be used to describe laser beam projected from air, which has refraction index $n_1 = 1$, to water having $n_2 = 1.33$, implying that $\theta_2 < \theta_1$ and the refracted ray would be closer to the normal line.

To study how the changes in the index of refraction and the randomness of the medium affect the laser light, some researchers concentrate the randomness into rectangular sheets called phase screens as shown in Figure 6.4. It is assumed that there is no randomness in between phase screens and turbulence from eddies are grouped in the nearby phase screen.

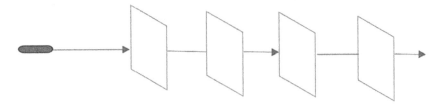

FIGURE 6.4
Phase Screens.

In some studies, researchers prefer to concentrate all the randomness in the medium in a single phase screen and often place it in the middle of the medium. In most cases, however, multiple thin phase screens are employed, where each phase screen indicates the change in the phase of the light wave, hence the name phase screen is used. On the other hand, thick phase screens account for fluctuations in the amplitude as well as in phase. Here we consider thin multiple phase screens as are more commonly used. In this setting, as the laser light propagates, meaning travels, in the z direction, it is likely that it enters mediums with varying refractive indices, which alter the phase, φ of the light wave and phase screens at these points on the z axis serve as indicators of the places of these phase fluctuations. The majority of work in the literature in this area has been on uncorrelated phase screens. Laser beams are usually assumed to be Gaussian and we note that for Gaussian random variables, uncorrelated implies independence. Therefore, the statistical properties as well as the temperature and intensity distributions of one phase screen should not affect those of any other phase screen and hence, sufficient distance is required to be placed in between the phase screens. In this chapter we explain an article regarding correlated phase screens.

Throughout the chapter we have used the material in [1,9,14,19] to define the terms needed and we recommend these sources for more information in the area. To make the material accessible to readers with a mathematical background, we have used the expected value, $\mathbb{E}(X)$ to denote the average mean of the random variable X instead of the notation, $< X >$ often used in physics and in most of the papers discussed here. To improve the presentation, we have denoted vector $\vec{x}$ as x, where x being a vector is made clear based on the context. We also let K be a positive constant in inequalities whose value does not have an affect on the estimates and may be kept as K from line to line.

6.1 Schrödinger Equation Modeling Laser Propagation

The laser light is viewed as an electromagnetic wave propagating along the horizontal z axis, where, as indicated in Figure 6.1, the vertical direction represents the electric field and the remaining axis is used to measure the magnetic field. Recall that the Schrödinger equation in its simplest form is, $iu_t + \Delta u = 0$, which has a similar form to the heat equation; however, unlike the heat equation, because of its complex nature, the Schrödinger equation does not posses smoothing effects. Denoting the amplitude in the electric field direction as $\psi(x, y, z, t)$, the following Schrödiner type equation is used to model the laser beam propagating in random media in the positive z direction,

$$2i\kappa \frac{\partial \psi}{\partial z} + \nabla_T^2 \psi + 2\kappa^2 n_1 \psi = 0, \tag{6.2}$$

where, $n_1 = n - 1$ is the random part of the refractive index, n, and for wavelength, λ, the constant, $\kappa := \frac{2\pi}{\lambda}$ is the wave number. The operator, $\nabla_T^2 := \frac{\partial^2}{\partial x^2} + \frac{\partial^2}{\partial y^2}$ is called the transverse Laplacian, since it is the sum on only the transverse variables, x, y and not on z. We note that ψ and n_1 depend on variables x, y, z, t, where in (6.2) we have suppressed the notation for better presentation.

Since the laser beam is viewed as an electromagnetic wave, equation (6.2) is derived from Maxwell's equations, where we refer to Sections 1.1–1.3 in [6] for details on the steps. It can also be shown that $\psi = \psi_0 e^{i\varphi}$ is a solution to (6.2), where ψ_0 is the initial amplitude at the transmitter and $\varphi := \varphi(x, y, z, t)$ is the phase of the light wave.

Another commonly studied model is

$$2i\kappa \frac{\partial \psi}{\partial t} + \nabla_T^2 \psi + v|\psi|^{2\sigma}\psi = 0, \tag{6.3}$$

$$\psi(x, 0) = \varphi(x),$$

for $\varphi \in L^2(\mathbb{R}^d)$ and $x \in \mathbb{R}^d$, where $\nabla_T^2 = \frac{\partial^2}{\partial x_1^2} + \frac{\partial^2}{\partial x_2^2} + \dots + \frac{\partial^2}{\partial x_d^2}$. In the case of $\sigma = 1$ and transverse dimension, $d = 2$, system (6.3) characterizes the propagation of intense laser beam in Kerr medium. Kerr nonlinearity refers to the nonlinearity, $|\psi|^2\psi$ and Kerr medium is any isotropic medium. A medium is said to be isotropic if values of the amplitude of the beam do not change depending on the direction of the propagation in the medium. If $v > 0$, the nonlinear Schrödinger equation (NLS) given by (6.3) is called focusing and when $v < -1$ it is called defocusing. For simplicity, many authors choose $v = 1$ or $v = -1$ for the two cases. Furthermore, terms subcritical, supercritical, and critical NLS are given to (6.3) with $\sigma d < 2$, $\sigma d > 2$, and $\sigma d = 2$, respectively. It is known that in the case of subcritical focusing, solutions exist globally and in the case of supercritical NLS, singular solutions may develop.

In the study of Schrödinger equations, inequalities called Strichartz estimates are commonly applied. To introduce them, we need the mild solution of the Schrödiner equation, which has the semigroup, $U(t) = e^{it\Delta}$. For instance, the mild solution of (6.3) is,

$$\psi(t) = U(t)\varphi + iv \int_0^t U(t - s)|\psi(s)|^{2\sigma}\psi(s)ds, \tag{6.4}$$

As we discuss in Chapter 8, mild solution is similar to weak solution, with the test function being the appropriate semigroup.

For the dimension denoted by d, two constants q and r are referred to as an admissible pair if

$$\frac{2}{q} = \frac{d}{2} - \frac{d}{r},$$

and

$$\begin{cases} 2 \leq r \leq \infty & d = 1, 2, \\ 2 \leq r < \frac{2d}{d-2} & d > 2. \end{cases}$$

With the notation above, the Strichartz estimates are as follows.

i. If $\varphi \in L^2(\mathbb{R}^\infty)$ and (q, r) is an admissible pair, then $U(t)\varphi \in L^q(\mathbb{R}, L^r(\mathbb{R}^d)) \cap C(\mathbb{R}, L^2(\mathbb{R}^d))$ and

$$\|U(\cdot)\varphi\|_{L^q(\mathbb{R}, L^r(\mathbb{R}^d))} \leq K\|\varphi\|_{L^2}. \tag{6.5}$$

ii. For interval $I \subset \mathbb{R}$ such that $t_0 \in \bar{I}$, if (q, r) and (γ, ρ) are admissible pairs and γ' and ρ' are the conjugates of γ and ρ, respectively, then for $f \in L^{\gamma'}(I, L^{\rho'})$,

$$\left\| \int_{t_0}^t U(t-s)f(s)ds \right\|_{L^q(I,L^r)} \leq K\|f\|_{L^{\gamma'}(I,L^{\rho'})}. \tag{6.6}$$

To illustrate the use of these estimates in an example, we explain the result in the following theorem. For its proof, notice that given two real-valued functions, u_1, u_2, which without loss of generality satisfy, $u_1 \leq u_2$, we have for any $\alpha > 0$,

$$\left| |u_1|^\alpha u_1 - |u_2|^\alpha u_2 \right| = \left| |u_1|^\alpha u_1 - |u_2|^\alpha u_1 + |u_2|^\alpha u_1 - |u_2|^\alpha u_2 \right|$$
$$\leq \left| |u_1|^\alpha u_1 - |u_2|^\alpha u_1 \right| + |u_2|^\alpha |u_1 - u_2|.$$

Now using the assumption, $u_1 \leq u_2 \leq |u_2|$,

$$\leq \left| |u_1|^\alpha u_2 - u_1^\alpha u_1 \right| + |u_2|^\alpha |u_1 - u_2| = \left| u_1^\alpha u_1 - |u_1|^\alpha u_2 \right| + |u_2|^\alpha |u_1 - u_2|$$
$$\leq |u_1|^\alpha |u_1 - u_2| + |u_2|^\alpha |u_1 - u_2| \leq \left(|u_1|^\alpha + |u_2|^\alpha \right) |u_1 - u_2|.$$

Thus,

$$\left| |u_1|^\alpha u_1 - |u_2|^\alpha u_2 \right| \leq |u_1 - u_2| \left(|u_1|^\alpha + |u_2|^\alpha \right). \tag{6.7}$$

Theorem 6.1 (*Proposition 4.2.1 and Lemma 4.2.2 in* [3]) *For $\varphi \in H^1(\mathbb{R}^N)$ and $\lambda \in \mathbb{C}$, if there exists a local mild solution to,*

$$iu_t + \Delta u + \lambda |u|^\alpha u = 0, \tag{6.8}$$
$$u(0) = \varphi,$$

in $H^1(\mathbb{R}^N)$, where $0 < \alpha < \infty$ if $N = 1,2$ and $0 < \alpha < \frac{4}{N-2}$ if $N > 2$, then the solution is unique.

Proof. The mild solution of (6.8) is given by,

$$u(t) = U(t)u_0 + i\lambda \int_0^t U(t-s)|u(s)|^\alpha u(s)ds.$$

Let the time interval be denoted by I, which is bounded by the fact that the solution is local and we may assume that I contains zero. Suppose u_1 and u_2 are two mild solutions of (6.8), then by (6.7) and Hölder's inequality for $r = \alpha + 2$,

$$\left\| \,|u_1|^\alpha\, u_1 - |u_2|^\alpha\, u_2 \,\right\|_{L^{r'}(I)} \leq K \| \, u_1 - u_2 \, \|_{L^r(I)} \left(\|u_1\|_{L^r(I)}^\alpha + \|u_2\|_{L^r(I)}^\alpha \right).$$

To apply the Strichartz estimates, we find the admissible pair, (r,q) by,

$$\frac{2}{q} = \frac{N}{2} - \frac{N}{\alpha+2} = \frac{N\alpha}{2(\alpha+2)},$$

giving $q = \frac{4(\alpha+2)}{N\alpha}$, which in terms of r is $q = \frac{4r}{N(r-2)}$. Let $J \subset I$ be a smaller interval that still contains zero, then the Hölder's inequality may be invoked on time to obtain,

$$\left\| \,|u_1|^\alpha\, u_1 - |u_2|^\alpha\, u_2 \,\right\|_{L^{q'}(J,L^{r'})}$$
$$\leq K_T \|u_1 - u_2\|_{L^{q'}(J,L^r)} \left(\|u_1\|_{L^\infty(J,L^r)}^\alpha + \|u_2\|_{L^\infty(J,L^r)}^\alpha \right). \tag{6.9}$$

Now by Strichartz estimate with admissible pair (q,r) for both (q,r) and (γ,ρ) in (6.6), along with (6.9) we have,

$$\|u_1 - u_2\|_{L^q(J,L^r)} = \left\| i\lambda \int_0^t U(t-s)\left(|u_1(s)|^\alpha u_1(s) - |u_2(s)|^\alpha u_2(s)\right)ds \right\|_{L^q(J,L^r)}$$
$$\leq K \left\| \lambda \left(|u_1(s)|^\alpha u_1(s) - |u_2(s)|^\alpha u_2(s) \right) \right\|_{L^{q'}(J,L^{r'})}$$
$$\leq K\|u_1 - u_2\|_{L^{q'}(J,L^r)} \left(\|u_1\|_{L^\infty(J,L^r)}^\alpha + \|u_2\|_{L^\infty(J,L^r)}^\alpha \right).$$

Since we are assuming that u_1 and u_2 are in $H^1(\mathbb{R}^N)$, we arrive at,

$$\|u_1 - u_2\|_{L^q(J,L^r)} \le K\|u_1 - u_2\|_{L^{q'}(J,L^r)}. \tag{6.10}$$

Next the following statement is proved to show that inequality (6.10) implies,

$$\|u_1 - u_2\|_{L^q(J,L^r)} = 0.$$

Let I be an interval that contains zero. If $1 \le a < b \le \infty$, then for every interval $J \subset I$ that also contains zero and for every $\phi \in L^b(I)$ that satisfies,

$$\|\phi\|_{L^b(J)} \le K\|\phi\|_{L^a(J)}, \tag{6.11}$$

we have $\phi = 0$ a.e. on J.

As is commonly reasoned in the literature, it is sufficient to prove the statement for $I = [0, T]$, since then the same proof may be adjusted to show the result for $[-T, 0]$ and then for all $T > 0$. Hence, let $T < \infty$ and $I = [0, T]$. We break up interval I by,

$$I = [0, \tau) \cup [\tau, t) \cup [t, T] = J_1 \cup J_2 \cup J_3,$$

where $0 \le \tau < T$ and $\tau < t < T$. We assume that on J_1, $\phi = 0$ a.e. and find by (6.11) and then Hölder's inequality,

$$\|\phi\|_{L^b(J_2)} \le K\|\phi\|_{L^a(J_2)} \le K(t-\tau)^{\frac{1}{a}-\frac{1}{b}}\|\phi\|_{L^b(J_2)}. \tag{6.12}$$

which gives, $\|\phi\|_{L^b(J_2)}\left(1 - K(t-\tau)^{\frac{1}{a}-\frac{1}{b}}\right) \le 0$. Now the value of τ may be changed to ensure that $K(t-\tau)^{\frac{1}{a}-\frac{1}{b}} < 1$ in order to yield, $\|\phi\|_{L^b(J_2)} = 0$, since the norm can not be negative. Thus, $\|\phi\|_{L^b(J_1 \cup J_2)} = 0$. As for interval J_3, let,

$$\theta = \sup\left\{0 < t < T : \|\phi\|_{L^b(0,t)} = 0\right\}.$$

We need to show that $\theta = T$. Suppose by contradiction, $\theta < T$, then $\tau = \theta$ with $\phi = 0$ a.e. on $(0, \theta)$. By above, $\phi = 0$ a.e. also on $(0, t)$, where $0 < t < T$ but this contradicts θ being the maximum time, t before which $\|\phi\|_{L^b(0,t)} = 0$ a.e.

Now with $a = q'$ and $b = q$, we have, $\|u_1 - u_2\|_{L^q(J,L^r)} = 0$. $\qquad\square$

Many researchers have studied the solutions of NLS equation (6.3) or small variations of the equation and have emphasized its importance in optics in the study of the laser beam propagation. For more background and proofs of the classical well-posedness results on linear and nonlinear Schrödinger equations, including the form in (6.13), we recommend [3] and for more advanced study in the area, we refer the reader to [6,21]. Here we focus on more recent results on solutions to biharmonic NLS given by,

$$i\psi_t(x,t) + \varepsilon_1 \Delta^2 \psi + |\psi|^{2\sigma} \psi = 0, \tag{6.13}$$

in the Sobolev space H^2, where $\Delta^2 \psi$ is the biharmonic operator and $\varepsilon_1 \in \{-1,1\}$. For a description on spaces used in this section see the Appendix at the end of the chapter.

In [13], B. Pausader considers the following biharmonic NLS equation with $u_0 \in H^2$,

$$i\partial_t u + \Delta^2 u + |u|^2 u = 0, \tag{6.14}$$

which is (6.13) with $\varepsilon_1 = \sigma = 1$. He studies its global well-posedness by applying his earlier results in [12]. Here E denotes the energy in the system.

Theorem 6.2 *(Theorem 5.1 in [13]) If $E_{max} < \infty$, there exists a maximal solution, $u \in \mathcal{C}(I, \dot{H}^2)$. In addition, for a function, g, defined as,*

$$g(h, x_0)u_0 = h^2 u_0(h(x - x_0)), \tag{6.15}$$

satisfying for every u_0, h, t_0, x_0, the set, $K = \{g(t)u(t) : t \in I\}$ is precompact in $\dot{H}^2$ and exactly one of the following three cases holds,

a. *$I = (0, \infty)$ and $|u(t)|^2 u(t) = t^{1/4}$ for all t,*
b. *$I = \mathbb{R}$ and $|u(t)|^2 u(t) = 1$ for all t,*
c. *$|u(t)|^2 u(t) \leq 1$ and $\liminf_{t \to \bar{T}} |u(t)|^2 u(t) = 0$, where $\bar{T} := \sup_{t \in I} t$, for all t.*

Maximal solutions are a form of local solutions with a precise definition given in the introduction of Chapter 8. In short, $u(t)$ is a maximal solution if there exists an increasing sequence of stopping times, $(\tau_n)_n$ satisfying, $\lim_{n \to \infty} \tau_n = \tau$ almost everywhere with $\tau \leq T$ and for all $n \geq 1$, $\sup_{t \in [0, \tau_n]} \|u(t)\|_{H^2} \geq n$. The solutions satisfying each of the cases in Theorem 6.2 are referred to as self-similar, soliton-like, and double low to high cascade, respectively. In the following result, notations $X \lesssim Y$ and $X \lesssim_\lambda Y$ are used to denote, $X \leq KY$ and $X \leq K(\lambda)Y$, respectively.

Theorem 6.3 *(Theorem 1.1 in [13]) For the biharmonic NLS (6.14), assuming that $u_0 \in H^2$ and $x \in \mathbb{R}^n$, the following hold,*

i. *if $1 \leq n \leq 8$, there exists a global solution, $u \in \mathcal{C}(\mathbb{R}; H^2)$,*
ii. *if $n \geq 9$, the solution blows up in H^2,*
iii. *if $5 \leq n \leq 8$, scattering occurs in H^2.*

Proof. To achieve statement (*i*), the author first concentrates on the energy-critical case, $n = 8$ and notes that a similar argument may be applied to extend it to all n in $1 \leq n \leq 8$. In Corollary 5.1 of the paper, the author proves that if $E(u_0) \leq E$, then any maximal solution, $u \in \mathcal{C}(I, \dot{H}^2)$ is a global solution of (6.14) and $\|u\|_{\dot{S}^2(\mathbb{R})} \leq C(E)$, where,

$$
\|u\|_{\dot{S}^s(I)} := \sup_{(a,b)} \left(\sum_{N=1}^{\infty} N^{2s+\frac{4}{a}} \|P_N u\|_{L^a(I,L^b)}^2 \right)^{1/2} \lesssim_s \sup_{(a,b)} \left(\sum_{N=1}^{\infty} N^{2s+\frac{4}{a}} \|u\|_{L^a(I,L^b)}^2 \right)^{1/2}.
$$

B. Pausader proves Corollary 5.1 mentioned above by a proof by contradiction. That is, suppose $\|u\|_{\dot{S}^2(\mathbb{R})} > C(E)$. Applying Strichartz-type estimates established in [12], the following inequality is derived,

$$
0 < C(E) < \|u\|_{\dot{S}^2(I)} \lesssim \|u_0\|_{\dot{H}^2} + \left\| |\nabla|^{2-\frac{2}{a}} u|u|^2 \right\|_{L^{a'}(I,L^{b'})}. \tag{6.16}
$$

Since it is assumed that u is a maximal solution, then $C(E)$ has to be finite. Noting that $C(E)$ depends on E, then the requirement, $E_{\max} < \infty$ is obtained.

Using $h(t) := |u(t)|^2 u(t)$ in (6.15), the author proves that in each of the three cases in Theorem 6.2, if the set, $K = \{g(t)u(t) : t \in I\}$ is precompact in $\dot{H}^2$, then $u(t) = 0$ for $t \in I \neq \mathbb{R}$. Thus, this leads to a contradiction, since each maximal solution, $u(t)$ takes on one of the three cases and based on inequality (6.16) cannot be zero in I.

For statement (*ii*), the author proves that in dimensions $n \geq 9$, for any fixed $\varepsilon > 0$, there exists a solution $u \in \mathcal{C}([0, \varepsilon]; \dot{H}^2)$ such that for some $0 < t_\varepsilon < \varepsilon$,

$$
\|u(0)\|_{\dot{H}^2} < \varepsilon, \quad \text{and} \quad \|u(t_\varepsilon)\|_{\dot{H}^2} > \frac{1}{\varepsilon}, \tag{6.17}
$$

and hence the solution forms a singularity, which in the literature is referred to as the solution having a blow up. In Lemma 4.1 of [13], it is shown that for $k > \frac{n}{2}$, there exists a unique solution $w^\nu \in \mathcal{C}([-T, T], H^k)$ for

$$
i\partial_t w + \nu^4 \Delta^2 w + |w|^2 w = 0, \tag{6.18}
$$

$$
w^\nu(0) = \phi,
$$

where ϕ is in the Schwartz space, $\nu \in (0, 1)$, and $T = c|\log \nu|^c$ for some $c > 0$. Letting $u = \lambda^2 w^\nu(\lambda^4 t, \lambda \nu x)$ in (6.14) and using (6.18), we obtain,

$$
u^{(\nu,\lambda)}(t, x) = \lambda^2 w^\nu(\lambda^4 t, \lambda \nu x), \tag{6.19}
$$

$$
u^{(\nu,\lambda)}(0, x) = \lambda^2 \phi(\lambda \nu x),
$$

is a solution to (6.14) for any $\lambda \in (0, \infty)$. To prove the first inequality in (6.17), using the norm of H^s defined in the Appendix along with a change of variables, we have that if $\lambda v \geq 1$,

$$\|u^{(v,\lambda)}(0)\|_{H^2}^2 = \frac{\lambda^4 (\lambda v)^{-2n}}{(2\pi)^n} \int_{\mathbb{R}^n} \left| \hat{\phi} \left(\frac{\xi}{\lambda v} \right) \right|^2 \left(1 + |\xi|^2 \right)^2 d\xi \lesssim_\phi \lambda^{8-n} v^{4-n} = \varepsilon^2.$$

Thus, setting $\lambda = \left(\varepsilon^2 v^{n-4} \right)^{\frac{-1}{n-8}}$ gives $\|u^{(v,\lambda)}(0)\|_{H^2}^2 \lesssim \varepsilon^2$.

For the second inequality in (6.17), observe that for $\tilde{w}(t) := \phi(x)e^{i|\phi(x)|^2 t}$,

$$i\partial_t \tilde{w} + |\tilde{w}|^2 \tilde{w} = i\phi(x) \left(i|\phi(x)|^2 \right) e^{i|\phi(x)|^2 t} + \left| \phi(x)e^{i|\phi(x)|^2 t} \right|^2 \phi(x)e^{i|\phi(x)|^2 t} = 0,$$

noting that $|e^{ix}| = 1$. Hence, $w^0(t, x) = \phi(x)e^{i|\phi(x)|^2 t}$ is a solution of Equation (6.18) with $v = 0$. Also the author verifies the following inequality for $|t| \leq c |\log v|^c$,

$$\|w^v(t)\|_{H^2} \gtrsim_\phi t^2.$$

Then with $t_v := c|\log v|^c$ we find, by the definition of $u^{(v,\lambda)}(t, x)$,

$$\left\| u^{(v,\lambda)}(\lambda^{-4} t_v) \right\|_{H^2}^2 = \left\| \lambda^2 (\lambda v)^{2-\frac{n}{2}} w^v(t_v) \right\|_{H^2}^2 \geq \left\| u^{(v,\lambda)}(\lambda^{-4} t_v) \right\|_{H^2}^2$$

$$\geq \lambda^{8-n} v^{4-n} \|w^v(t_v)\|_{H^2}^2 = \varepsilon^2 \left\| w^v(t_v) \right\|_{H^2}^2 \gtrsim_\phi \varepsilon^2 t_v^4.$$

Now letting $v \in (0, 1)$ be sufficiently small leads to sufficiently large t_v to obtain the second inequality for a given $\varepsilon > 0$.

For the proof of statement (*iii*), we refer the reader to the definition of scattering given in the Appendix. Here in the biharmonic case, the semigroup is $e^{it\Delta^2}$. In Proposition 11.2 of the paper, B. Pausader proves that in dimensions $5 \leq n \leq 8$, for any $u_+, u_- \in H^2$, there exists a unique solution of (6.14), $u \in \mathcal{C}(\mathbb{R}; H^2)$, satisfying,

$$\lim_{t \to \pm\infty} \|u(t) - e^{it\Delta^2} u_\pm\| = 0. \tag{6.20}$$

For the proof, the fixed point method is implemented by letting,

$$\Phi(u)(t) = e^{it\Delta^2} u_+ + i \int_t^\infty e^{i(t-s)\Delta^2} |u(s)|^2 u(s) ds, \tag{6.21}$$

for a given u_+, and $u \in \dot{S}^0(I) \cap \dot{S}^2(I)$, such that for sufficiently small $\delta > 0$, Φ is a contraction mapping on the set,

$$X_{T_\delta} = \left\{ u \in \dot{S}^0(I) \cap \dot{S}^2(I) : \|\nabla u\|_{L^{\frac{n+4}{2}}\left(I, L^{\frac{n(n+4)}{3n+4}}\right)} \le 2\delta \right.$$

$$\left. \text{and } \|u\|_{\dot{S}^0(I)} + \|u\|_{\dot{S}^2(I)} \lesssim \|u_+\|_{H^2} \right\},$$

under the norm $\dot{S}^0(I)$. Then Φ as a fixed point u and since $5 \le n \le 8$, then by statement (i), u is a global solution of (6.14) with,

$$u(t) = e^{it\Delta^2} u(t) + i \int_t^\infty e^{i(t-s)\Delta^2} |u(s)|^2 u(s) ds.$$

Hence,

$$\|u(t) - e^{it\Delta^2} u_+\|_{H^2} = \left\| i \int_t^\infty e^{i(t-s)\Delta^2} |u(s)|^2 u(s) ds \right\|_{H^2},$$

which goes to zero as $t \to \infty$. □

In Section Four of [7], G. Fibich, B. Ilan and G. Papanicolaou prove the global existence of solutions to (6.13) by letting $\tilde{\psi}(x,t) := e^{i\lambda^4 t} R_{B,\lambda}(x)$ with $\tilde{\psi}(x,0) = \varphi(x) \in L^2(\mathbb{R}^d)$, where $R_{B,\lambda}$ is a solution to,

$$-\lambda^4 R_{B,\lambda} + \varepsilon_2 \Delta^2 R_{B,\lambda} + R_{B,\lambda}^{2\sigma+1} = 0, \tag{6.22}$$

such that $\varepsilon_2 \in \{-1,1\}$. For the solution, $\psi(x,t)$ of (6.13), the Hamiltonian is defined as,

$$H(\psi) := -\varepsilon_1 \|\Delta\psi\|_2^2 - \frac{1}{\sigma+1} \|\psi\|_{2\sigma+2}^{2\sigma+2}, \tag{6.23}$$

and the conservation of Power and Hamiltonian imply,

$$\|\psi\|_2 = \|\varphi\|_2, \quad H(\psi) = H(\varphi). \tag{6.24}$$

For their result, we recall the Poincaré's and Gagliardo-Nirenberg inequalities as follows. Denoting $p^* = \frac{dp}{d-p}$, if $u \in W_0^{1,p}(U)$, where $1 \le p < d$ and U is a bounded subset in $\mathbb{R}^d$, then for every $1 \le q \le p^*$ the Poincaré's inequality gives,

$$\|u\|_{L^q(U)} \le K\|\nabla u\|_{L^p(U)}. \tag{6.25}$$

In the case $q = p = 2 < d$, with the help of integration by parts and the Cauchy-Schwarz inequality, the above leads to,

$$\|u\|^2_{L^2(U)} \le K\|\nabla u\|^2_{L^2(U)} \le K\int |\nabla u||\nabla u|dx = K\|\Delta u\|_{L^2(U)}\|u\|_{L^2(U)}. \tag{6.26}$$

There are different versions of Gagliardo-Nirenberg inequalities available in the literature. We will use the following original form in [11]. In spaces E, such as $\mathbb{R}$, if $u \in L_q(\mathbb{R}^d)$ where $1 \le q \le \infty$ and its derivatives of order m are in $L^r(\mathbb{R}^d)$, where $1 \le r \le \infty$, then if $0 \le j < m$,

$$\|D^j u\|_{L^p} \le K\|D^m u\|^a_{L^r}\|u\|^{1-a}_{L^q}, \tag{6.27}$$

such that a satisfies $\frac{j}{m} \le a \le 1$ and

$$\frac{1}{p} = \frac{j}{d} + a\left(\frac{1}{r} - \frac{m}{d}\right) + (1-a)\frac{1}{q}. \tag{6.28}$$

We note that there are special cases for inequality (6.27) as listed in [11], however, the results discussed here do not fall in those categories and inequality (6.27) may be applied directly.

Theorem 6.4 (*Lemma 4.1 and Theorem 4.2 in* [7]) $\tilde{\psi}(x,t) := e^{i\lambda^4 t}R_{B,\lambda}(x)$ *is a global nontrivial solution to the biharmonic NLS* (6.13) *in space* H^2 *if the following conditions hold,*

    i. $\varepsilon_2 < 0$ and $\sigma < \frac{4}{4-d}$ and,

    ii. $\varepsilon_1 > 0$ or,

        $\varepsilon_1 < 0$ and $\sigma d < 4$ or,

        $\varepsilon_1 < 0$ and $\sigma d = 4$ and $\|\varphi\|^2_2 \le \left(\frac{\sigma+1}{B_{\sigma,d}}\right)^{1/\sigma}$, where $B_{\sigma,d}$ is a constant depending on σ, d.

Proof. First we verify that (6.22) has a nontrivial solution relying on condition (*i*). Two new equations may be formed by multiplying Equation (6.22) by $R_{B,\lambda}$ and $x \cdot \nabla R_{B,\lambda}$, respectively, and then integrating both equations with respect to x to obtain,

$$-\lambda^4\int R^2_{B,\lambda}dx + \varepsilon_2\int R_{B,\lambda}\Delta^2 R_{B,\lambda}dx + \int R^{2\sigma+2}_{B,\lambda}dx = 0,$$

$$-\lambda^4\int x \cdot \nabla R_{B,\lambda}R_{B,\lambda}dx + \varepsilon_2\int x \cdot \nabla R_{B,\lambda}\Delta^2 R_{B,\lambda}dx + \int x \cdot \nabla R_{B,\lambda}R^{2\sigma+1}_{B,\lambda} = 0.$$

By the integration by parts, the above equations simplify to,

$$-\lambda^4 \|R_{B,\lambda}\|_2^2 + \varepsilon_2 \|\Delta R_{B,\lambda}\|_2^2 + \|R_{B,\lambda}\|_{2\sigma+2}^{2\sigma+2} = 0, \tag{6.29}$$

$$\lambda^4 \|R_{B,\lambda}\|_2^2 + \varepsilon_2 \left(\frac{4}{d} - 1\right) \|\Delta R_{B,\lambda}\|_2^2 - \frac{1}{\sigma+1}\|R_{B,\lambda}\|_{2\sigma+2}^{2\sigma+2} = 0. \tag{6.30}$$

If we multiply (6.30) by $(\sigma + 1)$ and add it to (6.29), we arrive at,

$$\left(-\lambda^4 + \lambda^4(\sigma+1)\right) \|R_{B,\lambda}\|_2^2 + \left(\varepsilon_2\left(\frac{4}{d} - 1\right)(\sigma+1) + \varepsilon_2\right)\|\Delta R_{B,\lambda}\|_2^2$$

$$:= \lambda^4 \sigma \|R_{B,\lambda}\|_2^2 + a_1 \|\Delta R_{B,\lambda}\|_2^2 = 0.$$

Also if we multiply (6.29) by $\left(\frac{4}{d} - 1\right)$ and subtract from it Equation (6.30), we obtain,

$$-\lambda^4 \frac{4}{d}\|R_{B,\lambda}\|_2^2 + \left(\left(\frac{4}{d} - 1\right) + \frac{1}{\sigma+1}\right)\|R_{B,\lambda}\|_{2\sigma+2}^{2\sigma+2} = 0,$$

which leads to,

$$-\lambda^4 \frac{4}{d}\|R_{B,\lambda}\|_2^2 + \frac{4}{d}\left(\frac{4\sigma+4-d\sigma}{4(\sigma+1)}\right)\|R_{B,\lambda}\|_{2\sigma+2}^{2\sigma+2}$$

$$= -\lambda^4 \frac{4}{d}\|R_{B,\lambda}\|_2^2 + \frac{4}{d}a_2\|R_{B,\lambda}\|_{2\sigma+2}^{2\sigma+2} = 0.$$

Thus, from the above two equations we obtain,

$$\lambda^4 \sigma \|R_{B,\lambda}\|_2^2 = -a_1 \|\Delta R_{B,\lambda}\|_2^2, \tag{6.31}$$

$$\lambda^4 \|R_{B,\lambda}\|_2^2 = a_2 \|R_{B,\lambda}\|_{2\sigma+2}^{2\sigma+2}. \tag{6.32}$$

To ensure that the sign of each side of equations (6.31) and (6.32) agree, we need $-a_1$ and a_2 to be positive, which are confirmed by condition (*i*) in the theorem.

By condition (*i*), we may let $\varepsilon_2 = -1$ in (6.22) for the remaining of the proof. In order to prove that $\widetilde{\psi}(x)$ is a global solution to (6.13), Strichartz-type estimates may be applied to first obtain local solutions in space H^2. Then to attain the global solution, it is shown that $\widetilde{\psi}(x, t)$ is uniformly bounded in H^2. More precisely, we need to verify that each case in condition (*ii*) implies that $\widetilde{\psi}, \nabla\widetilde{\psi}$, and $\Delta\widetilde{\psi}$ are in space $L^2(\mathbb{R}^d)$.

For $\varepsilon_1 > 0$, we let $\varepsilon_1 = 1$ and note that by (6.26) and (6.24),

$$\|\nabla\widetilde{\psi}\|_2^2 \leq \|\Delta\widetilde{\psi}\|_2\|\widetilde{\psi}\|_2 \leq K\|\Delta\widetilde{\psi}_2\|_2, \tag{6.33}$$

since, $\|\varphi\|_2 < \infty$. To show that $\Delta\widetilde{\psi} \in L^2(\mathbb{R}^d)$, we let $\varepsilon_1 = 1$ in $H(\widetilde{\psi})$ and find,

$$\|\Delta\widetilde{\psi}\|_2^2 = -H(\widetilde{\psi}) - \frac{1}{\sigma+1}\|\widetilde{\psi}\|_{2\sigma+2}^{2\sigma+2} < -H(\widetilde{\psi}) \leq |H(\widetilde{\psi})| = |H(\varphi)| < \infty,$$

where we have used the second equality in (6.24).

Now for $\varepsilon_1 < 0$, we let $\varepsilon_1 = -1$ and since $\|\widetilde{\psi}\|_2 = \|\varphi\|_2$ still holds true, based on (6.33), we need $\|\Delta\widetilde{\psi}\|_2 < \infty$. Here the Hamiltonian gives,

$$\|\Delta\widetilde{\psi}\|_2^2 = H(\widetilde{\psi}) + \frac{1}{\sigma+1}\|\widetilde{\psi}\|_{2\sigma+2}^{2\sigma+2} = H(\varphi) + \frac{1}{\sigma+1}\|\widetilde{\psi}\|_{2\sigma+2}^{2\sigma+2}. \qquad (6.34)$$

It is left to bound $\|\psi\|_{2\sigma+2}^{2\sigma+2}$, which may be done by applying the Gagliardo-Nirenberg inequality (6.27) with $j = 0$, $m = r = q = 2$ and $p = 2\sigma + 2$, which by using (6.28) give, $a = \frac{d\sigma}{2(2\sigma+2)}$ and we have,

$$\|\widetilde{\psi}\|_{2\sigma+2} \leq B_{\sigma,d}\|\Delta\widetilde{\psi}\|_2^{\frac{d\sigma}{2(2\sigma+2)}}\|\widetilde{\psi}\|_2^{\frac{4-\sigma(d-4)}{2(2\sigma+2)}}. \qquad (6.35)$$

Raising both sides of inequality (6.35) to power $2\sigma + 2$ and using it in (6.34) we obtain,

$$\|\Delta\widetilde{\psi}\|_2^2 \leq H(\varphi) + \frac{1}{\sigma+1}B_{\sigma,d}\|\Delta\widetilde{\psi}\|_2^{\frac{d\sigma}{2}}\|\varphi\|_2^{2\sigma+2-\frac{d\sigma}{2}}.$$

Since $\|\Delta\widetilde{\psi}\|_2^2$ appears on both sides of the inequality, we need $\frac{d\sigma}{2} < 2$ to ensure that $\|\Delta\widetilde{\psi}\|_2^2$ on the left hand side dominates or we can group the terms $\|\Delta\widetilde{\psi}\|_2^2$ on both sides by letting $\frac{d}{2} = 2$ and requiring its multiple on the right hand side to be less than one. That is , we need, $\frac{1}{\sigma+1}B_{\sigma,d}\|\Delta\widetilde{\psi}\|_2^{\frac{d\sigma}{2}} = \frac{1}{\sigma+1}B_{\sigma,d}\|\Delta\widetilde{\psi}\|_2^{2\sigma} \leq 1$. These two options imply the two assumptions in (*ii*) when $\varepsilon_1 = -1$.   □

6.2 Optimal Control

To determine where a function has a maximum or a minimum, recall that the function's derivative is set to zero and solved. For example, in economics, optimization problems provide information on how to maximize the profit or minimize the cost. For more involved situations, in which models are given by ordinary or partial differential equations, tools in the field of optimal control are often applied. This area of study provides the essential information different industries need to base their decisions and improve their outcomes. For instance, in cancer biology, such as in the paper [8] discussed here, researchers have applied optimal control theory to study how to minimize tumor growth. We begin in this section by giving some background on optimal control relying on the information in [16], and for examples explain results in [8]. Afterwards, we discuss some results on the optimal control of Schrödinger equations.

In optimal control theory, one studies the parameters in the model and sets up a cost functional, also referred to as the objective functional, denoted typically by $J(u)$ or $\Phi(u)$, that needs to be maximized or minimized. In most cases, the goal is to minimize the cost functional, $J(u)$, which is the same as maximizing $-J(u)$. To ensure that all criteria are met in the process, a set of admissible controls, usually denoted by $\mathcal{U}$ or V, is formed, where only solutions, u, from this set can be considered in the functional. In [8], minimizing tumor mass and the amount of drugs prescribed to treat a cancer patient are investigated. More precisely, their model for tumor volume, N, is given by,

$$\frac{dN}{dt} = rNF(N) - G(N,t),$$

where r and $F(N)$ are the tumor's growth rate and growth function, respectively and $G(N,t)$ represents the effects of the drug on the tumor, aimed to reduce the tumor's volume and hence is negative in above. Two objective functionals considered here to be minimized are,

$$J_\alpha(u) = \int_0^T \left(a(N - N_d)^2 + bu^2 \right) dt, \tag{6.36}$$

$$J_\beta(u) = aN(T) + b \int_0^T u(t)dt, \tag{6.37}$$

with admissible set of controls,

$$\mathcal{U} = \left\{ u \text{ measurable } : u(t) \geq 0 \text{ for all } t \in [0,T] \right\}, \tag{6.38}$$

and

$$V = \left\{ u \text{ measurable } : 0 \leq u(t) \leq M, \text{ for all } t \in [0,T] \right\}, \tag{6.39}$$

respectively, where a,b are positive constants. The first cost functional aims to minimize the difference between the actual tumor size, N, with the size desired, denoted by N_d. The purpose of the second cost functional is to minimize the negative effects of the drugs, where the first term characterizes the weakness of the body at the end of the treatment and the second term represents the effects of the toxic components of the drugs on other parts of the body.

The first step in problems in optimal control is to ensure that there exists a unique solution to the original equation, referred to as the state equation. Then the existence of an optimal function, u in the admissible set that minimizes or maximizes the cost functional is proved. Afterwards, the adjoint equation, as introduced below is formed and the existence and uniqueness of its solution are established. Moreover, the adjoint equation is used to obtain the necessary conditions on optimality, which are referred to as the characterization of optimal control.

The often applied, dynamic programming approach to optimal control uses the Hamilton-Jacobi-Hellman equation (HJB). To illustrate this method, suppose the state equation is,

$$x'(t) = f(t, u(t), x(t)), \qquad (6.40)$$
$$x(0) = x_0,$$

and the cost functional that we wish to maximize is,

$$J(u) = \int_0^T F(t, u(t), x(t))dt + S(x(T), T). \qquad (6.41)$$

The Hamiltonian equation for (6.40) and (6.41) is defined as,

$$H(x, u, p, t) = F(t, u, x) + f(x, u, t)p(t), \qquad (6.42)$$

where, $p(t)$ is called the adjoint equation of the optimal control problem. It is assumed that the adjoint equation satisfies the boundary condition, $p(T) = S_x(x(t), t)|_{x=x^*(T)}$, with the star indicating the optimal value of $x(t)$. We have,

$$x' = H_p \quad \text{and} \quad p' := -H_x, \qquad (6.43)$$

which by writing the second equation as $-dp = H_x dt$, integrating from t to T and using the boundary condition leads to,

$$p(t) = \int_t^T H_x(x(s), u(s), p(s), s)ds + S_x(x(t), t)|_{x=x^*(T)}.$$

In the case x and u are vectors, we use, $x' = \nabla_p H(x, u, p, s)$ and $p' = -\nabla_x H(x, u, p, s)$, where because H is a scalar function, are determined by, for example,

$$\nabla_x H = \begin{bmatrix} H_{x_1} \\ H_{x_2} \\ . \\ . \\ . \\ H_{x_n} \end{bmatrix}.$$

Solving the optimal control problem by the above setup and using equations in (6.43) is based on the widely applied Pontryagin's maximum principle (see Chapter 1 of [15] for the original description). This theory requires the optimal control, u^* and x^* to satisfy the Hamiltonian maximizing condition,

$$H(x^*(t), u^*(t), p(t), t) \geq H(x^*(t), u(t), p(t), t),$$

for all t in the given time interval and all u in the admissible set.

K. Fister and J. Panetta in [8] give examples of linear and nonlinear models in their study of the size of tumors and the effects of drugs on them. In Section Four in [8], they focus on the simplified linear versions of their model and writing $x = \ln N$, convert the cost functional $J_\beta(u)$ in (6.37) to,

$$J_1(u) = ae^{x(T)} + b \int_0^T u(t)dt, \tag{6.44}$$

to be minimized based on the admissible set V in (6.39). We will discuss their result on the optimal control of

$$\frac{dx}{dt} = -rx(t) - u(t)\frac{\delta}{k + e^{x(t)}}, \tag{6.45}$$

and note that the results on the other two linear models presented in the article follow the same reasoning. In above, r, k, δ are positive constants, and it is assumed that $0 < N_0 < 1$. The first step is to form the Hamiltonian by (6.42) and use it to determine the adjoint equation by (6.43). Here,

$$H(x, u, \psi_2, t) = bu(t) + \left(-rx(t) - u(t)\frac{\delta}{k + e^{x(t)}} \right) \psi_2(t), \tag{6.46}$$

where the solution to the adjoint equation is denoted by $\psi_2(t)$ and hence,

$$\frac{d\psi_2}{dt} = -\left(-r + u(t)\frac{\delta e^{x(t)}}{(k + e^{x(t)})^2} \right) \psi_2(t), \tag{6.47}$$

$$\psi_2(T) = \frac{d}{dx}\left(ae^{x(t)} \right)|_{x=x^*(T)} = ae^{x^*(T)}.$$

By the separation of variables technique in ordinary differential equations, we obtain,

$$\psi_2(t) = ae^{x(T)-\int_t^T \left(r-u(s)\frac{\delta e^{x(s)}}{(k+e^{x(s)})^2}\right)ds}.$$ (6.48)

To determine the Hamiltonian maximality condition, we find the derivative of the Hamiltonian with respect to u,

$$\frac{\partial H}{\partial u} = b - \frac{\delta}{k+e^{x(t)}}\psi_2(t),$$

and observe that if $b - \frac{\delta}{k+e^{x(t)}}\psi_2(t) < 0$, then H is decreasing with respect to u and since based on the admissible control set V, we have $u \in [0, M]$, then H attains its minimum in this case at the highest value obtained by u, that is at $u = M$. Similarly, we have $u = 0$ if $\frac{\partial H}{\partial u} > 0$.

When $\frac{\partial H}{\partial u} = 0$, then the control is said to be singular and we prove that this case is not possible by using contradiction. Suppose $\frac{\partial H}{\partial u} = 0$, then,

$$b - \frac{\delta}{k+e^{x(t)}}\psi_2(t) = 0,$$

which by taking the derivative with respect to time of both sides and using the expressions for $\frac{dx}{dt}$ and $\frac{d\psi_2}{dt}$ leads to,

$$\psi_2(t)r\left(k+e^{x(t)}+x(t)e^{x(t)}\right) = 0.$$

Since $a > 0$, then $\psi_2(t)$ given in (6.48) is positive. Thus,

$$(1+x(t))e^{x(t)}+k = 0,$$

making $x(t)$ be a constant. Noting that $\frac{dx}{dt} = f(x, u, t)$ in (6.42), we obtain $H(t, x, u, \psi_2) = bu$, which is minimized when $u(t)$ takes the least value possible, that is if $u(t) = 0$. However, it was concluded above that $u(t) = 0$ if $\frac{\partial H}{\partial u} > 0$ leading to a contradiction. Therefore,

$$u^*(t) = \begin{cases} 0 & b - \frac{\psi_2\delta}{k+e^{x(t)}} > 0, \\ M & b - \frac{\psi_2\delta}{k+e^{x(t)}} < 0, \end{cases}$$

is the optimal control for this problem and replacing u with u^* in equations for $\frac{dx}{dt}$ and $\frac{d\psi_2}{dt}$ give the corresponding optimal state and optimal adjoint equations.

The adjoint equation may also be defined by a similar function called the Lagrangian function given by,

$$L(x, u, p, \mu, t) = H(x, u, p, t) + \mu g(x, u, t),$$

where, $p' := -L_x(x, u, p, \mu, t)$. In [8], this form is used for the optimal control results of nonlinear models. For example, the Lagrangian function for the nonlinear model,

$$\frac{dN}{dt} = rN \ln \left(\frac{1}{N}\right) (1 - \delta u(t)),$$

subject to cost functional, $J_\alpha(u)$ given in (6.36) is,

$$L(N, u, \lambda_3, w_1) = a(N - N_d)^2 + bu^2 + \lambda_3 \left(rN \ln \left(\frac{1}{N}\right) (1 - \delta ru(t))\right) - w_1(t)u(t),$$

where, $w_1(t) \geq 0$ and λ_3 is the solution of the corresponding adjoint equation. By taking the derivative, $\frac{\partial L}{\partial u}$ and setting it to zero, the optimal control becomes,

$$u^*(t) = \frac{-\lambda_3 \delta rN \ln N + w_1}{2b}.$$

The goal in establishing an optimal control is to prove that there exists $h^* \in \mathcal{U}$ such that for all $h \in \mathcal{U}$,

$$J(h^*) \leq \inf_{h \in \mathcal{U}} J(h), \tag{6.49}$$

and that the minimum, $J(h^*)$ is attained. The common technique, as applied by the next results here, is to consider a minimizing sequence, $\{h_n\}_n \subset \mathcal{U}$ such that,

$$J(h^*) \leq \liminf_{n \to \infty} J(h_n), \tag{6.50}$$

and prove that the limit, $h^* = \lim_{n \to \infty} h_n$ exists and u^* is a solution to the state equation with control h^*. To verify that $J(h^*)$ has a minimum, one may prove its weak lower semi-continuity in order to apply the theorem below. A function $f : X \to Y$ is said to be weakly lower semi-continuous at $x \in X$ if for each sequence, $(x_n)_n$ in X such that x_n converges weakly to $x \in X$, we have, $f(x) \leq \liminf_{n \to \infty} f(x_n)$.

Theorem 6.5 (*Theorem 5.2.2 in [17]*) *Suppose $F : K \to \mathbb{R}$ is weakly lower semicontinuous at every point of K and K is closed and convex. If either K is bounded* *or F has the growth property at some point in K, then there exists $x_0 \in K$ such that for all $x \in K$, we have, $F(x_0) \leq F(x)$.*

Function $f : X \to Y$ has the growth property at $x \in X$, if there exists $y \in X$ and a positive constant k such that if $\|y - x\|_X > k$, then $f(y) > f(x)$. Thus, an increasing function has the growth property at every point.

We now recall the following two types of derivatives. A function $J : X \to Y$ is Gâteaux differentiable at $x \in X$ if for every $h \in X$,

$$\lim_{\alpha \to 0} \frac{J(x + \alpha h) - J(x)}{\alpha}, \tag{6.51}$$

exists and the limit, denoted by $\delta J(x)$, is the Gâteaux derivative of J at x. We note that in the literature, Gâteaux derivative is also referred to as the G-derivative or the first variation of J. The Fréchet derivative of J at $x \in X$, denoted as $J'(x)$ satisfies,

$$\lim_{h \to 0} \frac{J(x + h) - J(x) - J'(x)h}{\|h\|} = 0, \tag{6.52}$$

which may also be written as,

$$J(x + h) - J(x) = J'(x)h + o(\|h\|). \tag{6.53}$$

It may be shown that if a function J is Fréchet differentiable, then it is Gáteaux differentiable and the two derivatives become equivalent. In addition, Fréchet differentiability implies the continuity of the function. See Section 4.2 in [24] for proofs and more background on these types of derivatives.

Based on Theorem 6.5, to verify that the cost functional, $J(h)$ attains its minimum, many authors prove that $J(h)$ is Fréchet differentiable and apply the following fact.

Theorem 6.6 *(Corollary 5.2.4 in [17]) If $F : K \to \mathbb{R}$ is Gâteaux differentiable and convex, then it is a weakly lower semi-continuous function.*

Using Theorems 6.5 and 6.6, we now discuss results on the optimal control of Schödinger equation. In [2], L. Baudouin, O. Kavian, and J. Puel study the optimal control problem for

$$i\partial_t u + \Delta u + \frac{u}{|x - a(t)|} + V_1(x, t)u = 0, \tag{6.54}$$

$$u(x, 0) = u_0(x),$$

which is the Schrödinger equation with two real-valued potentials, $V_0 = \frac{1}{|x-a(t)|}$ and V_1, with $a(t) \in W^{2,1}(0, T)$. The domain is $\Omega = \{(x, t) : x \in \mathbb{R}^3, t \in (0, T)\}$ and the following spaces are used,

$$H_1 := \left\{ v \in L^2(\mathbb{R}^3) : \int_{\mathbb{R}^3} \left(1 + |x|^2\right) |v(x)|^2 dx < \infty \right\}, \tag{6.55}$$

$$H_2 := \left\{ v \in L^2(\mathbb{R}^3) : \int_{\mathbb{R}^3} \left(1 + |x|^2\right)^2 |v(x)|^2 dx < \infty \right\}. \tag{6.56}$$

In the first half of the article the authors prove the existence and uniqueness of solutions to (6.54) in $L^\infty(0, T; H^2 \cap H_2)$, where H^2 is the usual Sobolev space and establish the following estimate,

$$\|u\|_{L^\infty(0,T; H^2 \cap H_2)} + \|\partial_t u\|_{L^\infty(0,T; L^2)} \le C \|u_0\|_{H^2 \cap H_2}. \tag{6.57}$$

In the second half of the paper, they concentrate on the equation's optimal control by letting the cost functional depend on V_1 and be defined as,

$$J(V_1) = \frac{1}{2} \int_{\mathbb{R}^3} |u(x, T) - u_1(x)|^2 dx + \frac{r}{2} \|V_1\|_H^2, \tag{6.58}$$

with $u_1 \in L^2$. In a Hilbert space, W, imbedded in $W^{1,\infty}$, they provide the admissible control set as,

$$H = \left\{ V : \left(1 + |x|^2\right)^{-\frac{1}{2}} V \in H^1(0, T; W) \right\}. \tag{6.59}$$

For their proof, we need the following theorems that are widely applied in the literature. We denote the dual space of X as X' and recall that reflexive spaces, such as L^p spaces for $1 < p < \infty$, are those with $(X')' = X$. Also for a Banach space, X, a sequence, $(u_n)_n$ in X converges weakly if for all $v \in X'$ (called test function), $\langle u_n, v \rangle \to \langle u, v \rangle$ as $n \to \infty$.

Theorem 6.7 (*Proposition 10.13 in [24]*) *For a Banach space, X, and sequence $(u_n)_n$ in X,*

  i. *If every subsequence of $(u_n)_n$ has a further subsequence that converges strongly to u, then the original sequence, $(u_n)_n$, converges strongly to u. The same statement is true for weak convergence.*

  ii. *In a reflexive Banach space, every bounded sequence has a weakly convergent subsequence.*

iii. In a reflexive Banach space, if all the weakly convergent subsequences of a sequence have the same limit then the original sequence converges weakly to that limit.

It is also known that if w_n converges weakly to w, denoted by $w_n \rightharpoonup w$, then,

$$\|w\| \leq \liminf_{n \to \infty} \|w_n\|, \qquad (6.60)$$

(see for example Appendix D.4 in [4]). In addition, the following theorems are applied.

Theorem 6.8 *(Rellich-Kondrachov Compactness theorem, see for example Section 5.7 in [4]) If $U \subset \mathbb{R}^n$ is a bounded open set such that its boundary, ∂U is in $\mathcal{C}^1(\mathbb{R}^n)$, then $W^{1,p}(U) \subset L^q(U)$ is a compact embedding, where $1 \leq p < n$ and $1 \leq q < \frac{np}{n-p}$.*

Theorem 6.9 *(Theorem 5 in [18]) If for Banach spaces, X, B, Y, the embeddings,*

$$X \subset B \subset Y,$$

hold with $B \subset Y$ being a compact embedding and for $1 \leq p \leq \infty$,

    i. $(f_n)_n$ is bounded in $L^p(0, T; X)$ and

    ii. $\|f_n(t + h) - f_n\|_{L^p(0,T-h;Y)} \to 0$ as $h \to 0$ uniformly for each f_n,

then $(f_n)_n$ is relatively compact in $L^p(0, T; B)$ (and in $\mathcal{C}(0, T; B)$ if $p = \infty$).

Theorem 6.10 *(Theorem 10 in [2]) There exists a solution to the minimization problem given by (6.58) and (6.59). Furthermore, denoting the optimal control as V_1^* and considering the adjoint problem,*

$$i\partial_t p + \Delta p + \frac{p}{|x - a(t)|} + V_1^* p = 0, \qquad (6.61)$$
$$p(T) = u(T) - u_1,$$

in Ω, the following optimality condition holds for all $\tilde{V} \in H$.

$$r\langle V_1^*, \tilde{V}\rangle_H = \mathrm{Im} \int_0^T \int_{\mathbb{R}^3} \tilde{V}(x, t)u(x, t)\bar{p}(x, t)dxdt. \qquad (6.62)$$

Proof. To prove the existence of a minimizer, we need to show that there exists $V_1^* \in H$ with,

$$J(V_1^*) \leq \inf_{V_1 \in H} J(V_1). \tag{6.63}$$

Let $(V_1^n)_n$ be a sequence in H such that

$$\lim_{n \to \infty} J(V_1^n) = \inf_{V_1 \in H} J(V_1), \tag{6.64}$$

where,

$$J(V_1^n) = \frac{1}{2} \int_{\mathbb{R}^3} |u_n(x, T) - u_1(x)|^2 dx + \frac{r}{2} \|V_1^n\|_H^2, \tag{6.65}$$

and u_n satisfies,

$$i\partial_t u_n + \Delta u_n + \frac{u_n}{|x - a(t)|} + V_1^n u_n = 0, \tag{6.66}$$

$$u_n(0) = u_0,$$

then by the proof of the well-posedness of (6.54), Equation (6.66) has a unique solution. By (6.64) we have that $J(V_1^n)$ is convergent, which implies that it is bounded and using (6.65), we obtain $\|V_1^n\|_H < \infty$. Thus, by Theorem 6.7(ii) there exists a weakly convergent subsequence, which is still denoted as $(V_1^n)_n$ for better presentation and by (6.60) we obtain,

$$\|V_1^*\|_H \leq \liminf_{n \to \infty} \|V_1^n\|_H. \tag{6.67}$$

Note that based on (6.65), inequality (6.67) along with,

$$\|u(x, T) - u_1(x)\|_{L^2}^2 \leq \liminf_{n \to \infty} \|u_n(T, x) - u_1(x)\|_{L^2}^2, \tag{6.68}$$

we obtain our goal of $J(V_1^*) \leq \lim_{n \to \infty} J(V_1^n) = \inf_{V_1 \in H} J(V_1)$. Hence, we verify (6.68) by showing that $(u_n)_n$ converges to u in L^2. For this convergence, we show that $(u_n)_n$ is relatively compact in L^2. By (6.57), we have that $(u_n)_n$ is bounded in $L^\infty(0, T; H^2 \cap H_2)$ and its derivative is bounded in $L^\infty(0, T; L^2)$. Thus, based on Theorem 6.9, with $p = \infty$, $(u_n)_n$ is relatively compact in $\mathcal{C}(0, T; H^1 \cap H_1)$ if $H^2 \cap H_2 \subset H^1 \cap H_1 \subset L^2$, where the embedding $H^2 \cap H_2 \subset H^1 \cap H_1$ is compact.

To prove that $H^2 \cap H_2 \subset H^1 \cap H_1$ is compact, the authors follow the usual procedure of letting $(\varphi_n)_n$ be an arbitrary sequence in $H^2 \cap H_2$ that converges weakly to zero in $H^2 \cap H_2$ and prove that it also converges weakly to zero in $H^1 \cap H_1$. Since $\varphi_n \rightharpoonup 0$ in $H^2 \cap H_2$, then the sequence is bounded and for simplicity we let $\|\varphi_n\|_{H^2 \cap H_2} \leq 1$. We proceed to prove the weak convergence,

$\varphi_n \rightharpoonup 0$, in $H^1 \cap H_1$. For $R > 0$, letting $B(0, R)$ be a ball of radius R, we have by the definition of H_1 given in (6.55),

$$\|\varphi_n\|_{H_1} = \int_{\mathbb{R}^3} \left(1 + |x|^2\right) |\varphi_n(x)|^2 dx$$

$$\leq \int_{B(0,R)} (1 + R^2)|\varphi_n(x)|^2 dx + \int_{|x| \geq R} (1 + |x|^2) \frac{(1 + |x|^2)}{(1 + R^2)} |\varphi_n(x)|^2 dx$$

$$\leq (1 + R^2)\|\varphi_n\|_{L^2(B(0,R))} + \frac{1}{1 + R^2}\|\varphi_n\|_{H_2},$$

in which the second term on the right hand side drops by the assumption that $\varphi_n \rightharpoonup 0$ in $H^2 \cap H_2$. For the first term, as noted in Proposition 3.4 of [22], a special case of the Rellich-Kondrachov theorem is the compact embedding of $H^{s+\sigma}(M) \subset H^s(M)$ for any $s \in \mathbb{R}, \sigma > 0$ and compact set M. Thus, we obtain, $H^2(B(0, R)) \subset H^1(B(0, R))$, implying, $\|\varphi_n\|_{H^1(B(0,R))} \leq K\|\varphi_n\|_{H^2(B(0,R))}$. Then the assumption, $\varphi_n \rightharpoonup 0$ in H^2, yields $\varphi_n \rightharpoonup 0$ in $H^1(B(0, R))$, which by the definition of H^1 norm, gives, $\varphi_n \rightharpoonup 0$ in $L^2(B(0, R))$.

Now for the convergence in H^1, by the definition of the norm in H_2 given in (6.56), we find,

$$\|\varphi_n\|_{H^1}^2 \leq K\|\varphi_n\|_{L^2}\|\varphi_n\|_{H^2} \leq K\|\varphi_n\|_{L^2} = K\|\varphi_n\|_{L^2(B(0,R))} + K \int_{|x| \geq R} |\varphi_n(x)|^2 dx$$

$$\leq K\|\varphi_n\|_{L^2(B(0,R))} + K \int_{|x| \geq R} \frac{(1 + |x|^2)^2}{(1 + R^2)^2} |\varphi_n(x)|^2 dx$$

$$\leq K\|\varphi_n\|_{L^2(B(0,R))} + \frac{K}{(1 + R^2)^2}\|\varphi_n\|_{H_2}^2,$$

in which both terms on the right converge to zero. Therefore, the embedding $H^2 \cap H_2 \subset H^1 \cap H_1$ is compact and the same reasoning may be applied to prove the compact embedding of $H^1 \cap H_1$ in L^2.

To obtain (6.62), by the well-posedness result of (6.54), since $p(T) = u(T) - u_1 \in L^2$, (6.61) has a unique solution. In Lemma 13 of the paper, the authors prove that the functional $J(V_1)$ is differentiable with,

$$DJ(V_1)[\widetilde{V}_1] = z(T), \tag{6.69}$$

for $\widetilde{V}_1 \in H$, where, $z \in C([0, T]; L^2)$ is the unique solution of,

$$i\partial_t z + \Delta z + \frac{z}{|x - a(t)|} + V_1 z = -\widetilde{V}_1 u, \tag{6.70}$$

$$z(0) = 0.$$

Based on (6.58), the expression in (6.69) translates to,

$$\text{Re} \int_{\mathbb{R}^3} (u(x,T) - u_1(x))\overline{z(x,T)}dx + r < V_1, \tilde{V}_1 >_H = 0. \tag{6.71}$$

Note that by the integration by parts we have using $\bar{i} = -i$,

$$\int_0^T \int_{\mathbb{R}^3} i\partial_s z(x,s)\overline{p(x,s)}dxds = \int_0^T \int_{\mathbb{R}^3} z(x,s)\overline{i\partial_s p(x,s)}dxds$$

$$- \int_{\mathbb{R}^3} z(x,T)\overline{ip(x,T)}dxds, \tag{6.72}$$

and

$$\int_0^T \int_{\mathbb{R}^3} \Delta z(x,s)\overline{p(x,s)}dxds = \int_0^T \int_{\mathbb{R}^3} z(x,s)\overline{\Delta p(x,s)}dxds, \tag{6.73}$$

since $z(0) = 0$. Thus, if we multiply (6.70) by $\overline{p(x,t)}$, integrate with respect to x and t implementing (6.72) and (6.73) and then taking the imaginary parts of both sides, we arrive at,

$$\text{Im}\left[\int_0^T \int_{\mathbb{R}^3} \left(\overline{i\partial_s p(x,s)} + \overline{\Delta p(x,s)} + \overline{\left(\frac{1}{|x-a(t)|} + V_1 \right) p(x,s)} \right) z(x,s)dxds \right.$$

$$\left. - \int_{\mathbb{R}^3} z(x,T)\overline{ip(x,T)}dx \right] = -\text{Im} \int_0^T \int_{\mathbb{R}^3} \tilde{V}_1 u(x,s)\overline{p(x,s)}dxds, \tag{6.74}$$

noting that $V_0 = \frac{1}{|x-a(t)|}$ and V_1 are real-valued giving, $\overline{V_0 + V_1} = V_0 + V_1$. Observe that by (6.61) the group of terms in the first integral on the left hand side becomes zero. Also since by (6.61), $p(T) = u(T) - u_1$, we obtain with $\bar{i} = -i$,,

$$-\text{Im} \int_{\mathbb{R}^3} z(x,T)\overline{ip(x,T)}dx = \text{Im} i \int_{\mathbb{R}^3} z(x,T)\left(\overline{u(x,T) - u_1(x)} \right) dx$$

$$= \text{Re} \int_{\mathbb{R}^3} z(x,T)\left(\overline{u(x,T) - u_1(x)} \right) dx,$$

hence, (6.74) becomes,

$$\text{Re} \int_{\mathbb{R}^3} z(x,T)\left(\overline{u(x,T) - u_1(x)} \right) dx = -\text{Im} \int_0^T \int_{\mathbb{R}^3} \tilde{V}_1 u(x,s)\overline{p(x,s)}dxds.$$

Notice that for two complex numbers, z_1, z_2, we have, $\text{Re}(z_1\bar{z}_2) = \text{Re}(\bar{z}_1 z_2)$, thus, by (6.71) we obtain (6.62).   □

In [5], B. Feng, K. Wang and D. Zhao study the optimal control of equation,

$$iu_t + \Delta u = V(x)u + h(t)|u|^\alpha u, \tag{6.75}$$

$$u(0, x) = u_0,$$

in $\Omega = \{(x, t) : x \in \mathbb{R}^N \text{ for } N \leq 3, \ t \in [0, T]\}$, with cost functional,

$$J(h) = \langle u(T), Au(T) \rangle_{L^2}^2 + \gamma_1 \int_0^T \left(E'(t)\right)^2 dt + \gamma_2 \int_0^T (h'(t))^2 dt, \tag{6.76}$$

and admissible set,

$$C_h(R) = \left\{ h \in H^1(0, T) : |h(0)| \leq R \right\}, \tag{6.77}$$

for a given real number, $R > 0$. In above,

$$E'(t) = \frac{1}{\alpha + 2}h'(t) \int_{\mathbb{R}^N} |u(t, x)|^{\alpha+2} dx := \frac{1}{\alpha + 2}h'(t)w(t), \tag{6.78}$$

where, $0 < \alpha < \frac{4}{N}$ and γ_1, γ_2 are positive constants, $h : (0, T) \to \mathbb{R}$ is in $H^1(0, T)$ and $A : L^2(\mathbb{R}^N) \to L^2(\mathbb{R}^N)$ is a bounded, linear operator that is self-adjoint. Furthermore, $u_0 \in \Sigma$, where,

$$\Sigma := \left\{ u \in H^1 : xu \in L^2 \right\}, \tag{6.79}$$

with norm,

$$\|u\|_{\Sigma} = \|u\|_{W^{1,2}} + \|xu\|_{L^2}. \tag{6.80}$$

The potential, V, is assumed to be subquadratic, meaning that for all $k \in \mathbb{N}^N$, with $|k| \geq 2$, we have $V \in C^\infty(\mathbb{R}^N)$ and $\partial^k V \in L^\infty(\mathbb{R}^N, \mathbb{R})$. Before proving the optimal control of the solution of (6.75), the authors first establish the global existence and uniqueness of its solution and derive the following estimate,

$$\|xu(t)\|_{L^2}^2 + \|\nabla u(t)\|_{L^2}^2 \leq M_1 e^{M_2 t}, \tag{6.81}$$

for two positive constants, M_1 and M_2. For the proof of their result we state the following theorem.

Theorem 6.11 *(Proposition 21.35 (c,d) in [25]) For Banach spaces, X, Y, Z over $\mathbb{R}$ or $\mathbb{C}$,*

    i. *If the embedding $X \subset Y$ is continuous, then strong convergence of a sequence in X implies the strong convergence of that same sequence in Y with the same limit. The same statement is true for weak convergence.*

    ii. *If the embedding $X \subset Y$ is compact, then weak convergence in X implies strong convergence in Y with the same limit.*

Theorem 6.12 *(Theorems 1.1 and 1.2 in [5]) There exists a solution to the optimal control problem defined by (6.76)-(6.77). Furthermore, $J(h)$ is Fréchet differentiable in $H^1(0, T)$ with,*

$$J'(h)v = \int_0^T v(t)Re \int_{\mathbb{R}^N} \overline{v(t,x)} \left(|u|^\alpha u\right)(t,x)dxdt - 2\gamma_2 \int_0^T v'(t)h'(t)dt$$

$$- \frac{2\gamma_1}{(\alpha+2)^2} \int_0^T v'(t)h'(t)w(t)^2 dt, \tag{6.82}$$

where $v \in \mathcal{C}(0, T; \Sigma(\mathbb{R}^N))$ is the solution to the adjoint equation,

$$iv_t = -\Delta v + V(x)v + \frac{\alpha+2}{2}h(t)|u|^\alpha v + \frac{\alpha}{2}h(t)|u|^{\alpha-2}u^2\bar{v}$$

$$- \frac{2\gamma_1}{\alpha+2}(h'(t))^2 w(t)|u|^\alpha u,$$

$$v(T) = 4i \langle u(T), Au(T)\rangle_{L^2} Au(T). \tag{6.83}$$

Proof. Let $(h_n)_n \in C_h(R)$ be a minimizing sequence satisfying (6.50). By (6.81) and the fact that A is a bounded operator, we have by the form of the cost functional given in (6.76), $J(h_n) \leq K$ for every n. Observing that,

$$\int_0^t h'_n(s)ds = h_n(t) - h_n(0),$$

we have, since $h_n \in C_h(R)$,

$$h_n(t) \leq R + T^{\frac{1}{2}} \left(\int_0^T |h'_n(s)|^2 ds\right)^{\frac{1}{2}}, \tag{6.84}$$

by the Cauchy-Schwarz inequality. Since $J(h_n)$ is bounded, then by (6.76), $\int_0^T |h'_n(t)|^2 dt \leq K$, which by noting (6.84) and applying the Cauchy-Schwarz inequality to $\|h_n\|_{L^2}$, we obtain that both h_n and h'_n are in L^2 and hence $(h_n)_n$ is uniformly bounded in $H^1(0, T)$. By Theorem 6.7(ii) there exists a subsequence, still denoted by $(h_n)_n$, such that $h_n \rightharpoonup h_R^*$ in $H^1(0,T)$. Now for solutions, $(u_n)_n$, corresponding to the controls, $(h_n)_n$, we apply estimate (6.81) in (6.80) to obtain,

$$\|u_n\|_{L^\infty(0,T;\Sigma)} = T \max_{t\in[0,T]} \|u_n(t)\|_\Sigma \le K. \tag{6.85}$$

By the embedding, $L^\infty \subset L^2$, we arrive at, $\|u_n\|_{L^2(0,T;\Sigma)} \le K$ and hence there exists a subsequence, still denoted by $(u_n)_n$, such that $u_n \rightharpoonup u^*$ in $L^2(0,T;\Sigma)$. It is known that the embedding, $\Sigma \subset L^p$ is compact for $2 \le p < \frac{2N}{N-2}$, which by Theorem 6.11(ii) yields, $u_n \to u^*$ in $L^2(0,T;L^p)$ for $2 \le p < \frac{2N}{N-2}$. Since for every n, (6.85) holds and in above we obtained a subsequence of u_n that converges to u^* in Σ, then we may conclude that $u^* \in L^\infty(0,T;\Sigma)$. Moreover, based on the proof of the well-posedness of (6.75), the authors verify that u^* is a solution to (6.75) with control h^* and further show that $u_n \to u^*$ in $\mathcal{C}((0,T);L^2) \cap \mathcal{C}((0,T);L^{\alpha+2})$.

Now to ensure that $J(h_R^*) \le \inf_{h\in C_h(R)} J(h)$, we need, based on (6.50),

$$0 \le \liminf_{n\to\infty} \left(\langle u_n(T), Au_n(T)\rangle_{L^2}^2 - \langle u^*(T), Au^*(T)\rangle_{L^2}^2 \right)$$
$$+ \frac{\gamma_1}{\alpha+2} \liminf_{n\to\infty} \int_0^T \left(|h_n'(t)|^2 |w_n(t)|^2 - |h_R^*(t)|^2 |w^*(t)|^2 \right) dt$$
$$+ \gamma_2 \liminf_{n\to\infty} \int_0^T \left(|h_n'(t)|^2 - |(h_R^*)'(t)|^2 \right) dt$$
$$= I_1 + I_2 + I_3. \tag{6.86}$$

Note that $I_1 = I_3 = 0$ based on the convergence $u_n \to u^*$ in $L^2(0,T;\Sigma)$, $h_n \rightharpoonup h_R^*$ in $H^1(0,T)$ as shown above and the fact that A is a bounded operator, implying that $Au_n(T)$ has a subsequence that converges weakly to $Au^*(T)$. As for I_2, we make the observation that,

$$I_2 = \frac{\gamma_1}{\alpha+2} \liminf_{n\to\infty} \int_0^T \left(|w_n(t)|^2 - |w^*(t)|^2 \right) |h_n'(t)|^2 dt$$
$$+ \liminf_{n\to\infty} \int_0^T \left(|w^*(t)|^2 \left(|h_n'(t)|^2 - |(h_R^*)'(t)|^2 \right) \right) dt.$$

Using the fact that $f(x) = |x|^{\alpha+2}$ is a convex function, we obtain,

$$\left| w_n(t) - w^*(t) \right| \le \left| \int_{\mathbb{R}^N} |u_n(x,t)|^{\alpha+2} - |u^*(x,t)|^{\alpha+2} dx \right|$$
$$\le \int_{\mathbb{R}^N} \left| |u_n(x,t)| - |u^*(x,t)| \right|^{\alpha+2} dx, \tag{6.87}$$

which converges to zero as $n \to \infty$, since $u_n \to u^*$ in $\mathcal{C}((0,T); L^{\alpha+2})$. Then both terms in the expression of I_2 above go to zero using $h_n \in H^1(0,T)$ and (6.50) and thus (6.86) holds.

For system (6.75), to determine the Fréchet derivative of $J(h)$, the authors first prove in Proposition 4.2 of the paper that for controls, $h, \tilde{h} \in H^1(0,T)$ and their corresponding solutions, $u, \tilde{u} \in \mathcal{C}((0,T); \sum)$, if there exists $\varepsilon > 0$ such that $\|\tilde{h} - h\|_{H^1(0,T)} < \varepsilon$, then for every admissible pair, (γ, ρ),

$$\|\tilde{u} - u\|_{L^\gamma(0,T; \sum^{1,\rho})} \leq K\|\tilde{h} - h\|_{H^1(0,T)}, \tag{6.88}$$

where $\sum^{1,\rho} := \{u \in W^{1,\rho} : xu \in L^\rho\}$. Then letting $\tilde{h}(t) = h(t) + \phi(t)$, where $\phi(t)$ is a positive function taking values in the neighborhood of zero, they find using (6.88),

$$J(\tilde{h}) - J(h) = 2\gamma_2 \int_0^T h'(t)\phi'(t)dt + o\left(\|\phi\|^2_{H^1(0,T)}\right)$$

$$+ \int_0^T \phi(t)\text{Re} \int_{\mathbb{R}^N} \bar{v}(t,x)|u|^\alpha u \, dx \, dt + o\left(\|\phi\|^2_{H^1(0,T)}\right)$$

$$+ \frac{2\gamma_1}{(\alpha+2)^2} \int_0^T \phi'(t)h'(t)w^2(t)dt + o\left(\|\phi\|^2_{H^1(0,T)}\right),$$

which by setting $\|\phi\|_{H^1(0,T)}$ tend to zero and applying the definition of Fréchet derivative given in (6.53), yields (6.82). □

M. Subasi considers the following system in [20],

$$i\frac{\partial\psi}{\partial t} + a_0\frac{\partial^2\psi}{\partial x^2} - v(x,t)\psi = f(x,t), \tag{6.89}$$

$$\psi(x,0) = \phi(x),$$

$$\frac{\partial\psi(0,t)}{\partial x} = g_0(t), \quad \frac{\partial\psi(\ell,t)}{\partial x} = g_1(t),$$

in the domain, $\Omega = \{(x,t) : x \in (0,\ell), \quad t \in (0,T)\}$, where $\psi := \psi(x,t;v)$, $a_0 > 0$, $\phi \in W^{2,3}(0,\ell)$, $f \in W^{2,1}(\Omega)$, and $g_0, g_1 \in L^2(0,T)$. We note that notation $L_2(0,T)$ is used in their paper to denote the usual $L^2(0,T)$ space and here we will use L^2 to avoid the confusion with the Hilbert-Schmidt space, defined in Chapter 8, that is commonly denoted by L_2 in the literature. The cost functional to be minimized is given by,

$$J_\alpha(v) = \alpha_0 \int_0^T |\psi(0,t) - f_0(t)|^2 dt + \alpha_1 \int_0^T |\psi(\ell,t) - f_1(t)|^2 dt + \alpha\|v - w\|^2_{L^2(\Omega)}, \tag{6.90}$$

where $\alpha_0, \alpha_1, \alpha$ are positive numbers with $\alpha_0 + \alpha_1 > 0$, and $f_0, f_1 \in W^{2,1}(\Omega)$. Moreover, the potential $v(x,t)$ is assumed to be in the admissible set,

$$V = \left\{ v \in W^{2,1}(\Omega) : \quad \text{for some } \beta_0, \beta_1, \beta_2 > 0, \quad \beta_0 \le v(x,t) \le \beta_1, \quad \left| \frac{\partial v}{\partial t} \right| \le \beta_2 \right\}. \tag{6.91}$$

The author applies the following theorem.

Theorem 6.13 *(Theorem 5.2.8 in [17]) If $F : K \to \mathbb{R}$ is a strictly convex function, then it attains a unique minimum point.*

A function $F : K \to \mathbb{R}$ is strictly convex if its domain is convex and for every $t \in (0,1)$ and $x, y \in K$ with $x \neq y$,

$$F\left(tx + (1-t)y\right) < tF(x) + (1-t)F(y). \tag{6.92}$$

Theorem 6.14 *(Section 3 in [20]) There exists a unique solution to the minimization problem given by (6.89)-(6.90).*

Proof. Using the fact that L^2 is a convex space, it is sufficient to prove that $J_\alpha(v)$ is continuous to obtain the uniqueness of a minimum by Theorem 6.13. More precisely, the continuity of $J_\alpha(v)$ implies that there exists a dense subset of L^2 in which inequality (6.92) is attained. For the existence of a minimum, since V is a bounded and closed set, the continuity of $J_\alpha(v)$ along with Weirestrass theorem ensures that the minimizing sequence in (6.50) converges to a limit, $J(h_*)$. Thus, the focus of the proof is to verify that the cost functional is continuous. Writing $J_\alpha(v)$ as,

$$J_\alpha(v) = J_0(v) + \alpha \|v - w\|^2_{L^2(\Omega)}, \tag{6.93}$$

note that it is enough to show the continuity of $J_0(v)$. We find,

$$|J_0(v + \Delta v) - J_0(v)|$$

$$= \alpha_0 \left(\int_0^T |\psi(0, t; v + \Delta v) - f_0(t)|^2 \, dt - \int_0^T |\psi(0, t; v) - f_0(t)|^2 \, dt \right)$$

$$+ \alpha_1 \left(\int_0^T |\psi(\ell, t; v + \Delta v) - f_1(t)|^2 \, dt - \int_0^T |\psi(\ell, t; v) - f_1(t)|^2 \, dt \right)$$

$$= I_{11} + I_{12} + I_{21} + I_{22}.$$

Notice that by letting,

$$\Delta\psi(x,t) := \psi(x,t;v+\Delta v) - \psi(x,t;v), \tag{6.94}$$

we obtain,

$$I_{11} = \alpha_0 \int_0^T |\psi(0,t;v+\Delta v) - f_0(t)|^2\,dt = \alpha_0 \int_0^T |\Delta\psi(0,t) + \psi(0,t;v) - f_0(t)|^2\,dt$$

$$= \alpha_0 \int_0^T |\Delta\psi(0,t)|^2\,dt + 2\alpha_0 \int_0^T |\Delta\psi(0,t)|\,|\psi(0,t;v) - f_0(t)|\,dt$$

$$= \alpha_0 \int_0^T |\psi(0,t;v) - f_0(t)|^2\,dt,$$

which leads to,

$$I_{11} + I_{12} = \alpha_0 \int_0^T |\Delta\psi(0,t)|^2\,dt + 2\alpha_0 \int_0^T |\Delta\psi(0,t)|\,|\psi(0,t;v) - f_0(t)|\,dt.$$

Hence, we have,

$$|J_0(v+\Delta v) - J_0(v)|$$

$$\leq \alpha_0 \int_0^T |\Delta\psi(0,t)|^2\,dt + 2\alpha_0 \int_0^T |\psi(0,t;v) - f_0(t)|\,|\Delta\psi(0,t)|\,dt$$

$$+ \alpha_1 \int_0^T |\Delta\psi(\ell,t)|^2\,dt + 2\alpha_1 \int_0^T |\psi(\ell,t;v) - f_1(t)|\,|\Delta\psi(\ell,t)|\,dt$$

$$\leq \alpha_0 \|\Delta\psi(0,t)\|_{L^2(0,T)}^2 + 2\alpha_0 \|\psi(0,t;v) - f_0(t)\|_{L^2(0,T)} \|\Delta\psi(0,t)\|_{L^2(0,T)}$$

$$+ \alpha_1 \|\Delta\psi(\ell,t)\|_{L^2(0,T)}^2 + 2\alpha_1 \|\psi(\ell,t;v) - f_1(t)\|_{L^2(0,T)} \|\Delta\psi(\ell,t)\|_{L^2(0,T)},$$

by the Cauchy-Schwarz inequality, referred in the paper as the Cauchy-Bunyakowsky inequality. To bound $\|\Delta\psi(x,t)\|_{L^2}^2$ by a term involving Δv, we find a system for $\Delta\psi(x,t)$ using (6.94) and (6.89) as follows,

$$i\frac{\partial\Delta\psi}{\partial t} + a_0\frac{\partial^2\Delta\psi}{\partial x^2} - (v+\Delta v)\Delta\psi = \Delta v\psi, \tag{6.95}$$

$$\Delta\psi(x,0) = \frac{\partial\Delta\psi(0,t)}{\partial x} = \frac{\partial\Delta\psi(\ell,t)}{\partial x} = 0,$$

With the test function, $\overline{\Delta\psi} \in W^{2,1}(\Omega)$, the generalized solution to (6.95) is,

$$\int_0^\ell \int_0^T \left(i\frac{\partial\Delta\psi}{\partial t} + a_0\frac{\partial^2\Delta\psi}{\partial x^2} - (v+\Delta v)\Delta\psi \right)\overline{\Delta\psi}\,dt\,dx = \int_0^\ell \int_0^T (\Delta v\psi)\overline{\Delta\psi}\,dt\,dx,$$

$$\tag{6.96}$$

which simplifies to,

$$\int_0^\ell \int_0^T \left(i\frac{\partial}{\partial t}|\Delta\psi|^2 + a_0\frac{\partial^2|\Delta\psi|^2}{\partial x^2} - (v+\Delta v)|\Delta\psi|^2 \right) dtdx$$
$$= \int_0^\ell \int_0^t (\Delta v\psi)\overline{\Delta\psi}dtdx. \tag{6.97}$$

Noting that v is not a complex function, the complex conjugate of (6.97) becomes,

$$\int_0^\ell \int_0^T \left(-i\frac{\partial}{\partial t}|\Delta\psi|^2 + a_0\frac{\partial^2|\Delta\psi|^2}{\partial x^2} - (v+\Delta v)|\Delta\psi|^2 \right) dtdx$$
$$= \mathrm{Re}\int_0^\ell \int_0^T (\Delta v\psi)\overline{\Delta\psi}dtdx - i\mathrm{Im}\int_0^\ell \int_0^T (\Delta v\psi)\overline{\Delta\psi}dtdx. \tag{6.98}$$

Now subtracting (6.98) from (6.97) yields,

$$\int_0^\ell \int_0^T 2i\frac{\partial}{\partial t}|\Delta\psi|^2 dtdx = 2i\int_0^\ell \int_0^T (\Delta v\psi)\overline{\Delta\psi}dtdx.$$

That is,

$$\int_0^\ell |\Delta\psi|^2\,dx = \int_0^\ell \int_0^T (\Delta v\psi)\overline{\Delta\psi}dtdx,$$

which by Young's inequality gives,

$$\|\Delta\psi\|_{L^2(0,\ell)}^2 \leq \frac{1}{2}\|\Delta v\psi\|_{L^2(\Omega)}^2 + \frac{1}{2}\int_0^T \|\Delta\psi\|_{L^2(0,\ell)}^2 dt$$
$$\leq \|\Delta v\psi\|_{L^2(\Omega)}^2 + \int_0^T \|\Delta\psi\|_{L^2(0,\ell)}^2 dt.$$

Now we apply the Gronwall's inequality (see the Appendix) to arrive at,

$$\|\Delta\psi\|_{L^2(0,\ell)}^2 \leq \|\Delta v\psi\|_{L^2(\Omega)}^2 + \int_0^T e^{t-s}\|\Delta v\psi\|_{L^2(\Omega)}^2 ds$$
$$\leq \|\Delta v\psi\|_{L^2(\Omega)}^2 \left(1 + \left(-e^{t-T} + e^t \right) \right)$$
$$\leq \left(1 + e^T \right) \|\Delta v\psi\|_{L^2(\Omega)}^2 \leq K \left(\max_{(x,t)\in\Omega} |\Delta v|^2 \right) \|\psi\|_{L^2(\Omega)}^2$$
$$= K\|\Delta v\|_{L^\infty(\Omega)}^2,$$

observing that $\psi \in L^2(\Omega)$. Since K does not depend on Δv, then setting $\|\Delta v\|_{L^\infty(\Omega)}$ to zero yields the result. ☐

6.3 Phase Screens

Here we introduce some terms and concepts that are often used in articles on optics and we apply them to explain a result on phase screens used to study the effects of temperature fluctuations on the phase of the laser beam as it propagates in random media. To visualize and examine the phase screens, numerical simulations are typically employed in which the process is discretized to develop a computer code. In these simulations, the discrete Fourier transform (DFT) is a common tool, which is given by,

$$\mathcal{F}_k = \frac{1}{\sqrt{2\pi}} \sum_{k=0}^{N-1} f_k \, e^{-\frac{2\pi i n k}{N}}, \tag{6.99}$$

where f_ks in above are sample points of function f. Recall that any periodic function, $f(t)$ with period T, may be represented by its Fourier transform, given in one dimension as,

$$\mathcal{F}(\omega) = \frac{1}{\sqrt{2\pi}} \int_{-\infty}^{\infty} f(t) \, e^{-i\omega t} dt, \tag{6.100}$$

with the inverse Fourier transform,

$$f(t) = \frac{1}{\sqrt{2\pi}} \int_{-\infty}^{\infty} \mathcal{F}(\omega) \, e^{i\omega t} d\omega, \tag{6.101}$$

where $\omega = \frac{2\pi}{T}$. Using the Dirac δ-function,

$$\delta(x - x_0) = \begin{cases} 0 & \text{if } x \neq x_i, \\ \infty & \text{if } x = x_i \end{cases}, \tag{6.102}$$

equation (6.101) may be written as,

$$\int f(x_i)\delta(x - x_i)dx_i = f(x). \tag{6.103}$$

From (6.102) one can observe that the Dirac δ-function is zero for all $x \neq 0$, has a sharp pulse to infinity at $x = 0$ and satisfies, $\int_{-\infty}^{\infty} \delta(x - x_i)dx_i = 1$. In higher dimensions, $n > 1$, the Fourier transform of $u \in L^1(\mathbb{R}^n)$ is given by,

$$\mathcal{F}u(y) = \frac{1}{(2\pi)^{n/2}} \int_{\mathbb{R}^n} e^{-ix\cdot y} u(x)dx,$$

with,

$$\mathcal{F}^{-1}u(y) = \frac{1}{(2\pi)^{n/2}} \int_{\mathbb{R}^n} e^{ix\cdot y} u(x)dx,$$

as its inverse.

To rigorously measure the strength of turbulence in the medium, the correlation function, structure function, and the refractive index spectral density are determined. For a random variable, $X(t)$, representing for example, wind velocity, temperature, pressure or the refractive index in the medium, the correlation function, also referred to as the auto-correlation function, is defined as,

$$B_X(t_1, t_2) = \mathbb{E}(\, X(t_1)\, \overline{X(t_2)}\,), \tag{6.104}$$

for times, $0 < t_1 < t_2$, where $\bar{z}$ is the complex conjugate of z. The structure function is,

$$D_X(t) = \mathbb{E}(\, |X(t_2) - X(t_1)|^2\,), \tag{6.105}$$

which can be represented in terms of the correlation function as follows,

$$D_X(t) = \mathbb{E}\left(\, (X(t_2) - X(t_1))\, (\overline{X(t_2)} - \overline{X(t_1)})\, \right)$$
$$= B_X(t_2, t_1) - B_X(t_1, t_2) - B_X(t_2, t_1) + B_X(t_1, t_1).$$

The refractive index spectral density, denoted by Φ_n, is the Fourier transform of the correlation function of the refractive index, n. There are three main models in the literature available to calculate Φ_n. Denoting the eddy size under observation by $\tilde{L}$, the most commonly used model is the Kolmogorov power spectrum given by,

$$\Phi_n(\kappa) = 0.033\, C_n^2\, \kappa^{-11/3}, \tag{6.106}$$

which can only be applied when $\kappa := \frac{2\pi}{\tilde{L}}$, called the scalar spatial frequency, is in the range $\frac{2\pi}{L_0} < \kappa < \frac{2\pi}{\ell_0}$. Note that based on the definition of κ, this condition is equivalent to the eddy being in the inertial subrange, $\ell_0 < \tilde{L} < L_0$.

In (6.106) it is assumed that $\Phi(\kappa) = 0$ for $\kappa > \frac{2\pi}{\ell_0}$ and the constant C_n^2 is called the index of refraction structure constant used to indicate how much the refractive index varies within a medium. In the case $\kappa > \frac{2\pi}{\ell_0}$, that is when the eddy size satisfies, $\tilde{L} < \ell_0$, the Tatarski spectrum is often applied and is defined as,

$$\Phi_n(\kappa) = 0.033 \, C_n^2 \, \kappa^{-11/3} \, \exp\left(-\frac{\kappa^2}{\kappa_m^2}\right), \tag{6.107}$$

where, $\kappa_m = \frac{5.91}{\ell_0}$. The third common model, called the Von Kármán spectrum, is given by,

$$\Phi_n(\kappa) = 0.033 \, C_n^2 \left(\kappa^2 + \frac{1}{L_0^2}\right)^{-11/6} \exp\left(-\frac{\kappa^2}{\kappa_m^2}\right), \tag{6.108}$$

and it may be used for any value of κ.

Phase screens are layers used in simulations to indicate the location of changes in the beam's phase. In [23] P. Paramonov, A. Vorontsov and V. Kunitsyn simulate three-dimensional correlated phase screens. They determine the correlation within each phase screen by calculating its auto-correlation and also find the correlation between two consecutive phase screens by using the cross-correlation function. Here correlation refers to the dependence of one phase screen on the other. The propagation path in the z direction is assumed to be from 0 to L and the screens are placed along the propagation axis, where the location of the jth phase screen is denoted by z_j. Then $\{z_j\}_{j=0}^{J}$ is the set of phase screens with $z_0 = 0$ and $z_J = L$. For a point $\zeta \in [0, L]$ on the propagation z axis, the fluctuation in the phase of the laser beam and hence phase screen is given by,

$$\tilde{\varphi}(x, y, \zeta) = \int_0^\zeta n(x, y, \zeta') d\zeta'. \tag{6.109}$$

Furthermore, the jth phase screen, $\tilde{\varphi}_j$ is modeled by,

$$\tilde{\varphi}_j(\rho) = \int e^{i(\kappa, \rho)} \sqrt{\Phi_{11}(\kappa)} \, Z_j(d^2\kappa), \tag{6.110}$$

where, Z_j is a random variable having cross-correlation of screens at z_{j_1} and z_{j_2} as follows,

$$B_{Z_{j_1}, Z_{j_2}} = \mathbb{E}\left(Z_{j_1}(d^2\kappa_1)\overline{Z_{j_2}(d^2\kappa_2)} \right) = \delta(\kappa_1 - \kappa_2) \, R_{j_1, j_2}\, (\kappa) d^2\kappa_1 d^2\kappa_2. \quad (6.111)$$

Φ_{11} in (6.110) is the Fourier transform of B_{11}, which is the autocorrelation of phase screen $\widetilde{\varphi}_1$. For two points $r_1 = (x_1, y_1, \zeta_1)$ and $r_2 = (x_2, y_2, \zeta_2)$ on the specific phase screen represented by $\widetilde{\varphi}(x, y, \zeta)$, the authors determine the auto-correlation as follows,

$$B_{\widetilde{\varphi}}(r_1, r_2) = \mathbb{E}(\widetilde{\varphi}(r_1)\overline{\widetilde{\varphi}(r_2)}) = \mathbb{E}\left(\int_0^{\zeta_1} n(x_1, y_1, \zeta_1')d\zeta_1' \int_0^{\zeta_2} n(x_2, y_2, \zeta_2')d\zeta_2' \right)$$

$$= \int_0^{\zeta_1} \int_0^{\zeta_2} \mathbb{E}\left(n(x_1, y_1, \zeta_1') n(x_2, y_2, \zeta_2') \right) d\zeta_1' d\zeta_2'$$

$$= \int_0^{\zeta_1} \int_0^{\zeta_2} B_n\left(x_2 - x_1,\ y_2 - y_1,\ \zeta_2' - \zeta_1' \right) d\zeta_1' d\zeta_2'.$$

It is known that, $B_n(r) = \mathcal{F}\Phi_n(r)$, where in this paper they use the Von Kármán power spectrum, $\Phi_n(\kappa)$ defined by (6.108). Moreover, by a classical theorem,

$$B_n(r) = \beta^2 \, G(\alpha r), \quad (6.112)$$

where the r in the left hand side is a vector, whereas on the right hand side is the scalar $|r|$. Recall that the gamma function is given by $\Gamma(\alpha) = \int_0^\infty x^{\alpha-1}e^{-x}dx$ for $\alpha > 0$. Using the gamma function, for $p \geq 0$,

$$G(p) := \Gamma\left(\frac{2}{3}\right) \sum_{m=0}^{\infty} \left(\frac{p}{2}\right)^{2m} \frac{1}{m!} \left(\frac{1}{\Gamma\left(m + \frac{2}{3}\right)} - \frac{\left(\frac{2}{3}\right)^{2/3}}{\Gamma\left(m + \frac{4}{3}\right)} \right), \quad (6.113)$$

for $p \geq 0$. Thus,

$$B_{\widetilde{\varphi}}(r_1, r_2) = \beta^2 \int_0^{\zeta_1} \int_0^{\zeta_2} G(\alpha r) \, d\zeta_1' d\zeta_2'.$$

To find the auto-correlation, which is correlation on the same phase screen, they let $\zeta_1 = \zeta_2 = z_j$ and $\rho = |(x_2, y_2) - (x_1, y_1)|$ and obtain after some estimations,

$$B_{\widetilde{\varphi}}(r_1, r_2) = B_{\widetilde{\varphi}}(\rho, z_j, z_j) \approx \beta^2 \, z_j^2 \, G(\alpha\rho).$$

As for cross-correlation between two phase screens, $\widetilde{\varphi}_{j_1}, \widetilde{\varphi}_{j_2}$, representation (6.110) is used to find,

$$B_{\widetilde{\varphi}_{j_1}, \widetilde{\varphi}_{j_2}}(\rho_1, \rho_2) = \mathbb{E}\left(\widetilde{\varphi}_{j_1}(\rho_1)\overline{\widetilde{\varphi}_{j_2}(\rho_2)}\right)$$

$$= \mathbb{E}\left(\int e^{i(\kappa_1, \rho_1)} \sqrt{\Phi_{11}(\kappa_1)}Z_{j_1}(d^2\kappa_1) \overline{\int e^{i(\kappa_2, \rho_2)} \sqrt{\Phi_{11}(\kappa_2)}Z_{j_2}(d^2\kappa_2)}\right)$$

$$= \int\int e^{i(\kappa_1, \rho_1)} e^{-i(\kappa_2, \rho_2)} \sqrt{\Phi_{11}(\kappa_1)\Phi_{11}(\kappa_2)} \, \mathbb{E}\left(Z_{j_1}(d^2\kappa_2)\overline{Z_{j_2}(d^2\kappa_2)}\right)$$

$$= \int\int e^{i(\kappa_1, \rho_1) - i(\kappa_2, \rho_2)} \sqrt{\Phi_{11}(\kappa_1)\Phi_{11}(\kappa_2)}\delta(\kappa_1 - \kappa_2) \, R_{j_1, j_2}(\kappa) \, d^2\kappa_1 d^2\kappa_2.$$

Letting $\kappa = |\kappa_2 - \kappa_1|$ and $\rho = |\rho_2 - \rho_1|$, we have by (6.103),

$$B_{\widetilde{\varphi}_{j_1}, \widetilde{\varphi}_{j_2}} = \int e^{i(\kappa, \rho)} \, \Phi_{11}(\kappa) \, R_{j_1, j_2}(\kappa) d^2\kappa,$$

leading to,

$$R_{j_1, j_2}(\kappa) = \frac{\mathcal{F}^{-1}(B_{\widetilde{\varphi}_{j_1}, \widetilde{\varphi}_{j_2}})}{\Phi_{11}(\kappa)}.$$

For simplicity, the authors let $j_1 = 1$ and $\widetilde{\varphi}_j$ be any phase screen afterwards and formulate the following expression for $B_{1,j}(\rho)$ in terms of $G(p)$ defined in (6.113).

$$B_{1,j}(\rho) = -\beta^2 \, (j-2) \, z_1 \int_{(j-2)z_1}^{(j-1)z_1} G\left(\alpha \sqrt{\rho^2 + u^2}\right) du$$

$$+ \beta^2 j z_1 \int_{(j-1)z_1}^{jz_1} G\left(\alpha \sqrt{\rho^2 + u^2}\right) du$$

$$+ \frac{\beta^2}{\alpha^2} \int_{k_1}^{\infty} u \, G(u) du - 2 \int_{k_2}^{\infty} u \, G(u) du + \int_{k_3}^{\infty} u \, G(u) du,$$

where,

$$k_1 = \alpha \sqrt{\rho^2 + (j-2)^2 \, z_1^2}, \quad k_2 = \alpha \sqrt{\rho^2 + (j-1)^2 \, z_1^2}, \quad k_3 = \alpha \sqrt{\rho^2 + j^2 \, z_1^2}.$$

To simulate the phase screens, they use the following presentation,

$$\tilde{\varphi}_j(x,y) = \int e^{i(\lambda x + \mu y)}\sqrt{\Phi_{11}(\lambda,\mu)}\, Z_j(d\lambda, d\mu), \qquad (6.114)$$

which is similar to (6.110). They view each phase screen as a square box with sides and thickness of measure $2q$. Namely, a phase screen is the three dimensional layer, $Q = [-q,q]_x \times [-q,q]_y \times [-q,q]_z$, where the axis of propagation runs through the center of the phase screen. To discretize (6.114), for $N \geq 1$ the screen is divided into equal size boxes of side length $d = \frac{q}{N}$. We refer to d as the spatial step size. Since Z_j is a random variable, we consider both spatial and spectral domains. The spatial domain as noted above is Q. For spectral domain we let $\Omega = [-\omega,\omega]_\lambda \times [-\omega,\omega]_\mu$ and denote the step size in Ω as $\delta = \frac{\omega}{N}$. Hence, for $m,n,k,\ell \in \{-N, ..., N-1\}$, (md, nd) gives the spatial node (x_m, y_n) and $(k\delta, \ell\delta)$ gives the spectral node (λ_k, μ_ℓ).

With the above notation, (6.114) may be presented in a sum form as follows,

$$\tilde{\varphi}_j(x_m, y_n) = \sum_{k=-N}^{N-1}\sum_{\ell=-N}^{N-1} e^{i\lambda_k x_m + i\mu_\ell y_n}\sqrt{\Phi_{11}(\lambda_k,\mu_\ell)}\, Z_j(d\lambda_k, d\mu_\ell)$$

$$= \sum_{k=-N}^{N-1}\sum_{\ell=-N}^{N-1} e^{i(k\delta)(md) + i(\ell\delta)(nd)}\sqrt{\Phi_{11}(\lambda_k,\mu_\ell)}\, Z_j(d\lambda_k, d\mu_\ell).$$

The authors then use the properties of Z_j and Φ_{11} to obtain,

$$\sqrt{\Phi_{11}(\lambda_k,\mu_\ell)}\, Z_j(d\lambda_k, d\mu_\ell) = \sqrt{\Phi_{11}^{k\ell}}\, Z_j^{k\ell},$$

yielding,

$$\tilde{\varphi}_j(x_m, y_n) = \tilde{\varphi}_j(md, nd) = \sum_{k=-N}^{N-1}\sum_{\ell=-N}^{N-1} e^{id\delta(km+\ell n)}\sqrt{\Phi_{11}^{k\ell}}\, Z_j^{k\ell}, \qquad (6.115)$$

which is the discrete Fourier transform of $\sqrt{\Phi_{11}^{k\ell}}\, Z_j^{k\ell}$. Using (6.115), the authors simulate two correlated phase screens in a random medium with total propagation length $L = 0.1m$ by letting $N = 512$ and $q = 0.02$ and obtain the following images in Figure 6.5, which is Figure 4 in their paper. As it may be observed from these simulations, the location of the dark concentrations in both screens verify the correlation between the two phase screens.

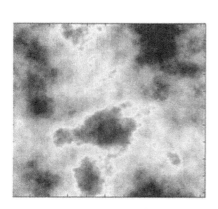
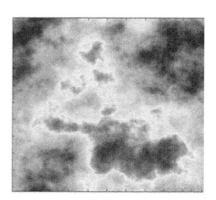

FIGURE 6.5
Simulation of Two Phase Screens.

6.4 Appendix

Spaces $\mathcal{S}(\mathbb{R}^n)$, $H^s(\mathbb{R}^n)$, $\dot{H}^s(\mathbb{R}^n)$ and $H^{-1}(\mathbb{R}^n)$ (see Section 1.4 in [3]):
The Schwartz space, $\mathcal{S}(\mathbb{R}^n)$, is the space of rapidly decreasing functions defined by,

$$\mathcal{S}(\mathbb{R}^n) = \left\{ \phi \in \mathcal{C}^\infty(\mathbb{R}^n) \; : \; \|\phi\|_{\alpha,\beta} < \infty \text{ for all } \alpha, \beta \in \mathbb{N} \cup \{0\} \right\},$$

with

$$\|\phi\|_{\alpha,\beta} = \sup_{|p| \leq \beta} \sup_{x \in \mathbb{R}^n} (1 + |x|)^\alpha \left| D^p \phi(x) \right|,$$

where, $p = (p_1, ..., p_n)$, $|p| = p_1 + ... + p_n$ and $D^\alpha \phi = \frac{\partial^{|\alpha|} \phi}{\partial x_1^{\alpha_1} ... \partial x_n^{\alpha_n}}$, and in dimension one it simplifies to,

$$S(\mathbb{R}) = \left\{ \phi \in \mathcal{C}^\infty(\mathbb{R}) : \|\phi\|_{\alpha,\beta} = \sup_{x \in \mathbb{R}} \left| x^\alpha \phi^{(\beta)}(x) \right| < \infty, \text{ for all } \alpha, \beta \in \mathbb{N} \right\}.$$

$\mathcal{S}(\mathbb{R}^n)$ is a Fréchet space and its dual, denoted by $\mathcal{S}'(\mathbb{R}^n)$ is called the space of tempered distributions.

Using the Schwartz space, we can define $H^s(\mathbb{R}^n)$, which is a different space than the widely used Sobolev space, $H^2(\mathbb{R}^n) = W^{2,2}(\mathbb{R}^n)$. Denoting the Fourier transform of u as $\hat{u}$, for $\xi \in \mathbb{R}^n$, and fixed $s \in \mathbb{R}$,

$$H^s(\mathbb{R}^n) := \left\{ u \in \mathcal{S}'(\mathbb{R}^n) : \left(1 + |\xi|^2\right)^{\frac{s}{2}} \hat{u} \in L^2(\mathbb{R}^n) \right\}, \qquad (6.116)$$

i.e. its norm is given by,

$$\|u\|_{H^s(\mathbb{R}^n)} = \left\| \left(1 + |\xi|^2\right)^{\frac{s}{2}} \hat{u} \right\|_{L^2(\mathbb{R}^n)},$$

where u is in $\mathcal{S}'(\mathbb{R}^n)$. The space, $\dot{H}^s(\mathbb{R}^n)$ is called the homogeneous Sobolev space and to define it let

$$\psi_j(\xi) = \eta\left(\frac{\xi}{2^j}\right) - \eta\left(\frac{\xi}{2^j - 1}\right),$$

where,

$$\eta(\xi) = \begin{cases} 1 & \text{if } |\xi| \le 1, \\ 0 & |\xi| \ge 2 \end{cases}$$

which makes $\phi_j \in S(\mathbb{R}^n)$ for each j. Then $\dot{H}^s(\mathbb{R}^n)$ is given by,

$$\dot{H}^s(\mathbb{R}^n) := \left\{ u \in \mathcal{S}'(\mathbb{R}^n) : \sum_{j=-\infty}^{\infty} \mathcal{F}^{-1}\left(|\xi|^s \psi_j \hat{u}\right) \in L^2(\mathbb{R}^n) \right\},$$

with norm,

$$\|u\|_{\dot{H}^s(\mathbb{R}^n)} = \left\| \sum_{j=-\infty}^{\infty} \mathcal{F}^{-1}\left(|\xi|^s \psi_j \hat{u}\right) \right\|_{L^2(\mathbb{R}^n)}. \qquad (6.117)$$

The more general spaces, $H^{s,p}(\mathbb{R}^n)$ and $\dot{H}^{s,p}(\mathbb{R}^n)$ are defined by (6.116) and (6.117), respectively with $L^p(\mathbb{R}^n)$ replacing $L^2(\mathbb{R}^n)$. That is,

$$H^{s,p}(\mathbb{R}^n) = \left\{ u \in \mathcal{S}'(\mathbb{R}^n) : \left(1 + |\xi|^2\right)^{\frac{s}{2}} \hat{u} \in L^p(\mathbb{R}^n) \right\},$$

with the convention that $H^s = H^{s,2}(\mathbb{R}^n)$.

For any space, $\mathcal{U}$, the notation $\mathcal{U}_0$ or $\mathcal{U}_c$ refers to the space with compact support. Furthermore, $W^{-m,p}$ denotes the dual space of $W^{m,p}$ and hence, H^{-1} is the dual of space H^1.

Scattering theory (see Chapter 7 in [3]):
For the NLS,

$$iu_t + \Delta u + f(u) = 0, \tag{6.118}$$

if the nonlinear term, $f(u)$, is zero, then (6.118) corresponds to the linear case with the unique solution, $u(t) = S(t)u(0)$, where $S(t) = e^{it\Delta}$. If a NLS equation has a global unique solution, u, then it is said to scatter if it takes the form of a linear Schrödinger equation as $t \to \infty$ and $t \to -\infty$. More precisely, there exist u_+ and u_- with,

$$u_+ = \lim_{t \to \infty} S(-t)u(t),$$

$$u_- = \lim_{t \to -\infty} S(-t)u(t),$$

such that with the associated semigroup of u denoted as $S(t)$, we have both of the following two limits,

$$\lim_{t \to \infty} \|u(t) - S(t)u_+\| = 0, \tag{6.119}$$

$$\lim_{t \to -\infty} \|u(t) - S(t)u_-\| = 0. \tag{6.120}$$

We note that in [3], the limit (6.119), similarly (6.120), is given as,

$$\lim_{t \to \infty} \|S(-t)u(t) - u_+\| = 0, \tag{6.121}$$

however, (6.121) is equivalent to (6.119) noting that

$$\|S(t)(S(-t)u(t) - u_+)\| = \|u(t) - S(t)u_+\|,$$

which follows from the properties of semigroup provided in Chapter 8.

Gronwall inequality (see for example Lemma 6.1.1 in [10]):
For functions, $f, g \in L^1[a, b]$,

$$f(t) \le g(t) + K \int_a^t f(s)ds, \tag{6.122}$$

implies,

$$f(t) \leq g(t) + K \int_a^t e^{K(t-s)} g(s) ds, \qquad (6.123)$$

which in the case that $g(t)$ is a constant, C, becomes,

$$f(t) \leq C e^{K(t-a)}.$$

References

1. L. Andrews, R. Phillips (2005). *Laser Beam Propagation through Random Media: Second Edition*. The International Society for Optical Engineering (SPIE), Bellingham, Washington.
2. L. Baudouin, O. Kavian, and J. Puel (2005). Regularity for a Schrödinger equation with singular potentials and application to bilinear optimal control. *J. Diff. Eq.* vol. 216, 188–222.
3. T. Cazenave (2003). *Semilinear Schrödinger Equations*. American Mathematical Society Courant Institute of Mathematical Sciences, vol. 10, New York, New York.
4. L. Evans (2010). *Partial Differential Equations: Second Edition*. American Mathematical Society, Graduate Studies in Mathematics, vol. 19, Providence, Rhode Island.
5. B. Feng, K. Wang and D. Zhao (2018). Optimal nonlinearity control of Schrödinger equation. *Evolution Eq. Contr. Theo.* vol. 7, no. 2, 317–334.
6. G. Fibich (2015). *The Nonlinear Schrödinger Equation: Singular Solutions and Optical Collapse*. Springer Applied Mathematical Sciences, vol. 192, New York.
7. G. Fibich, B. Ilan, and G. Papanicolaou (2002). Self-focusing with fourth-order dispersion. *SIAM J. Appl. Math.* vol. 62, no. 4, 1437–1462.
8. K. Fister and J. Panetta (2003). Optimal control applied to competing chemotherapeutic cell-kill strategies. *SIAM J. Appl. Math.* vol. 63, no. 6, 1954–1971.
9. A. Ishimaru (1997). *Wave Propagation and Scattering in Random Media*. IEEE/OUP Series on Electromagnetic Wave Theory. Oxford University Press, Oxford, UK.
10. G. Kallianpur, P. Sundar (2014). *Stochastic Analysis and Diffusion Processes*. Oxford Graduate Texts in Mathematics, vol. 24, Oxford University Press, Oxford, United Kingdom.
11. L. Nirenberg (1959). On elliptic partial differential equations. *Ann. Scuola Norm. Sup. Pisa.* vol. 13, no. 2, 115–162.
12. B. Pausader (2007). Global well-posedness for energy critical fourth-order Schrödinger equations in the radial case. *Dynamics of PDE.* vol. 4, no. 3, 197–225.
13. B. Pausader (2009). The cubic fourth-order Schrödinger equation. *J. Func. Anal.* vol. 256, 2473–2517.

14. F. Pedrotti, L. Pedrotti, and L. Pedrotti (2018). *Introduction to Optics: Third Edition.* Cambridge University Press, Cambridge, UK.

15. L. Pontryagin, V. Boltyanskii, R. Gamkrelidze and E. Mishchenko (1986).*The Mathematical Theory of Optimal Processes: Volume Four.* John Wiley & Sons Inc. New York.

16. S. Sethi (2019). *Optimal Control Theory: Applications to Management Science and Economics: Third Edition.* Springer, Switzerland.

17. R. Showalter (2010). *Hilbert Space Methods in Partial Differential Equations.* Dover Publications, Inc. Mineola, New York.

18. J. Simon (1987). Compact sets in the space $L^p(0, T; B)$. *Annali di Mathematica Pura ed Applicata.* vol. 146, no. 1, 65–96.

19. J. Strohbehn (Editor) (1978). *Laser Beam Propagation in the Atmosphere.* Topics in Applied Physics, vol. 25, Springer, Berlin, Germany.

20. M. Subasi (2002). An optimal control problem governed by the potential of a linear Schrödinger equation.*Appl. Math. Comp.* vol. 131, 95–106.

21. C. Sulem and P. Sulem (1999). *The Nonlinear Schrödinger Equation: Self-Focusing and Wave Collapse.* Springer Applied Sciences, vol. 139, New York.

22. M. Taylor (1996). *Partial Differential Equations I: Basic Theory.* Applied Mathematical Sciences, vol. 115, Springer, New York.

23. P. Paramonov, A. Vorontsov, and V. Kunitsyn (2015). A three-dimensional refractive index model for simulation of optical wave propagation in atmospheric turbulence.*Waves in Random and Complex Media.* vol. 25, no. 4, 556–575.

24. E. Zeidler (1986). *Nonlinear Functional Analysis and its Applications I: Fixed-Point Theorems.* Springer, New York, NY.

25. E. Zeidler (1990). *Nonlinear Functional Analysis and its Applications II/A: Linear Monotone Operators.* Springer, New York, NY.

7

PageRank in Google

The number of citations of a research article is an indicator of the article's importance and influence in the literature. This type of comparison is applied by Google to determine the order by which to list websites for a search inquiry. Each website has an importance score that ranks its popularity. This score is computed by an algorithm called PageRank. The greater the PageRank, the higher the website will appear on the list. It is intuitively clear that the developers prefer their websites to be listed toward the top to increase the chances of being viewed by surfers. Here we describe how researchers study links in the worldwide web, explain the power method commonly used to compute PageRank and discuss ways spammers use to manipulate their sites' PageRank.

The web is typically viewed as a directed graph with websites thought as vertices or nodes and links between them as edges. A link on a page i that can be used to go to page j is called an outgoing link for page i and an inlink for page j. Each edge connecting two sites is assigned a probability and these probabilities are formed into a large matrix, H, called the hyperlink matrix. Entries in H are determined by $h_{ij} = \frac{1}{\deg(P_i)}$, where $\deg(P_i)$, sometimes denoted as $|P_i|$, is the number of outlinks of webpage P_i. Then the PageRank of P_i is the sum of the PageRanks of all pages referring to it. More precisely, if B_{P_i} is the set of pages linked to P_i, then,

$$r(P_i) = \sum_{P_j \in B_{P_i}} \frac{r(P_j)}{\deg(P_j)}, \qquad (7.1)$$

gives the PageRank of page P_i. Thus, not having many outgoing links and being referred to by sites having high value PageRank increase the webpage's rank. In an attempt to enhance their PageRank, some web developers create irrelevant links and advertisements pointing to their sites, an act that is referred to as link spamming. In Section 7.3 we discuss different types of link spamming and how they can be detected.

Websites that have no outgoing links are called dangling nodes. For each dangling node, P_i, we assign $h_{ij} = 0$ for all j, making the row in H corresponding to a dangling node be a zero row. This makes the hyperlink matrix to be very sparse, meaning a matrix with many zero entries and thus can cause complications in the calculations of the PageRank. In Section 7.2 we describe

---

DOI: 10.1201/9781003299073-7

some results in the literature aimed to group dangling nodes to reduce their effects on computations.

To introduce the Google matrix used to compute the PageRank, we recall that a row stochastic matrix, also referred to as a stochastic matrix, has the property of the entries in each row summing to 1; whereas, a column stochastic matrix has entries in each of its columns add to 1. Furthermore, a probability vector is a vector whose entries sum to 1. The Google matrix is given by,

$$G(\alpha) = \alpha S + (1 - \alpha)ev^T, \tag{7.2}$$

where, $S : n \times n$ is a row stochastic matrix and vector v is a column probability vector, referred to as the personalization vector, since it records preferences of a generic surfer for each webpage. Vector e denotes an n-dimensional column vector having 1 for every entry. Hence, a probability vector, v, satisfies, $v^T e = 1$. Once a surfer is on a webpage, there are two ways he/she can leave it. One is to click on a link on the page and another is to jump to another webpage by closing the current webpage and using another URL or back button. For a random surfer, the probability of choosing the first option is denoted by α and the probability of the second option is $1 - \alpha$. The parameter $\alpha \in [0, 1]$ is called the damping factor for which Google typically assigns the fixed probability, $\alpha = 0.85$ for each webpage.

Recall that an eigenvector that solves the usual equation, $Ax = \lambda x$ is called a right eigenvector; whereas, a left eigenvector, y is one that satisfies, $y^*A = \lambda y^*$, where y^* denotes the conjugate transpose of y. Also for a matrix, A, the 1-norm is given by $\|A\|_1 = \max_j \sum_{i=1}^{p} |a_{ij}|$, which in the case of a nonzero probability vector, v, becomes $\|v\|_1 = 1$. PageRank is found to be the left hand eigenvector of the Google matrix, G, corresponding to the dominant eigenvalue, $\lambda = 1$. It is given by the probability vector, π^T, where each component, π_i, is the PageRank of webpage P_i. Therefore, to verify that a vector, π is the PageRank of the Google matrix, G, one needs to prove that $\pi \geq 0$, $\|\pi\|_1 = 1$ and $\pi^T G = \pi^T$.

The power method is widely applied in the literature to compute the PageRank. It is an iterative algorithm given by,

$$\pi_k^T = \pi_{k-1}^T G,$$

with the initial vector usually assumed to be $x_0^T = \frac{e^T}{n}$. The PageRank vector is then given by, $\pi^T = \lim_{k\to\infty} \pi_k^T$ if the limit exists. The rate of this convergence is equivalent to the rate of convergence of $\alpha^k \to 0$. Thus, the smaller the damping factor, the faster the power method converges. Some techniques have been proposed in the literature to increase the speed of this convergence (see for example, [8,12,19,24]). The power method and the idea of PageRank itself, were first introduced in 1998 by two researchers, S. Brin and L. Page, working at Google. Their research were presented in [9,10] and was originally based on a Markov chain model. We have provided a brief introduction to Markov chains in the Appendix and defined the terms and concepts needed for this chapter. For the information, we have mainly relied on [7,27] and refer to them for a more in depth study of Markov chains.

Since the hyperlink matrix, H is composed of probabilities with each entry giving the probability of going from one page to another, then it is closely related to a transition probability matrix. Furthermore, the likelihood of a surfer visiting a particular webpage does not depend on the webpages the surfer has visited in the past, making the process a Markov chain since there are large but finitely many webpages. In most cases, the matrix H does not meet the requirement of a transition probability matrix since some rows might not sum to 1. To convert H to a row stochastic matrix, the founders, S. Brin and L. Page, proposed to replace each zero row made by a dangling node with the vector $\frac{1}{n}e^T$. The new matrix is then denoted by $\bar{P}$. Afterwards, to ensure that the matrix is irreducible, they added the matrix, $E = \frac{ee^T}{n}$ to obtain,

$$\bar{\bar{P}} = \alpha\bar{P} + (1 - \alpha)E. \tag{7.3}$$

Comparing the above with (7.2), $\bar{\bar{P}} := G$, $\bar{P} := S$ and $E := ev^T$. Since the Markov chain based on the transition matrix, $\bar{\bar{P}}$, is irreducible, homogeneous and on a finite set of states, n, then it has a unique stationary distribution (see the Appendix). In this setting, to prove π is the PageRank vector, one has to show $\pi = \pi G$ and $\|\pi\|_1 = 1$, which is similar to the methods used to compute PageRank solely based on directed graphs. We first explain some results on the eigenvalues of the Google matrix in Section 7.1 and then in Section 7.2 we focus on ways to compute the PageRank based on both the directed graph and the Markov chain models. In Section 7.3, we discuss how PageRank can be optimized and ways some web developers attempt to increase their websites' rank. In this chapter, for background on PageRank we have used material in [20,23] and for more information on this area we refer the reader to [5,6,11,21].

7.1 Spectrum of the Google Matrix

We find the eigenvalues of the Google matrix based on the eigenvalues of matrix S in (7.2). The following is the Perron-Frobenius theorem which is commonly used in this context.

Theorem 7.1 *(See for example Theorem 9.34 in [25]) If matrix $A : n \times n$ is irreducible, nonnegative and primitive, then it has a positive real eigenvalue, λ_1 with algebraic multiplicity 1 such that $|\lambda_j| < |\lambda_1|$ for every other eigenvalue, λ_j.*

Since each entry in a stochastic matrix, A, is less than 1, then none of its eigenvalues can be greater than 1. Furthermore, A satisfies $Ae = e$, implying that $\lambda = 1$ is an eigenvalue of A with corresponding eigenvector, e. Therefore, $\lambda_1 = 1$ is the dominant eigenvalue. Since the Google matrix is stochastic, nonnegative, primitive as well as irreducible, then we obtain $\lambda_1 = 1 > |\lambda_2| \geq |\lambda_3| \geq ... \geq |\lambda_n|$.

For the next result, we recall similarity transformations. Suppose λ is an eigenvalue of a matrix $A : n \times n$ and $T : n \times n$ is an arbitrary invertible matrix. If we let vector y be defined as $y = T^{-1}x$, where x is the eigenvector associated with λ, then $Ty = x$ and

$$T^{-1}ATy = T^{-1}A(Ty) = T^{-1}Ax = T^{-1}(\lambda x) = \lambda T^{-1}x = \lambda y.$$

Thus, $T^{-1}ATy = \lambda y$, implying that λ is an eigenvalue of $T^{-1}AT$. Therefore, λ is an eigenvalue of A and $T^{-1}AT$. By the definition of similarity, A and T are similar and matrix T defined as above is called a similarity transformation.

Theorem 7.2 (Theorem 4.7.1 in [23]) *If the spectrum of S is $\{1, \lambda_2, ..., \lambda_n\}$, then the spectrum of the Google matrix is $\{1, \alpha\lambda_2, ..., \alpha\lambda_n\}$.*

Proof. The matrix S being a row stochastic matrix, gives $Se = e$, implying that e is an eigenvector of S associated with eigenvalue 1. For similarity transformation, we let, $Q = [e \quad X]$ where X is an arbitrary matrix and denote $(Q^{-1})^T = [y^T \quad Y^T]$. Then,

$$Q^{-1}\left(\alpha S + (1-\alpha)ev^T\right)Q = \alpha Q^{-1}SQ + (1-\alpha)Q^{-1}ev^TQ. \tag{7.4}$$

For the first term in (7.4), note that,

$$Q^{-1}Q = \begin{bmatrix} 1 & 0 \\ 0 & I \end{bmatrix} \quad \text{and} \quad Q^{-1}Q = \begin{bmatrix} y^T \\ Y^T \end{bmatrix}[e \quad X] = \begin{bmatrix} y^Te & y^TX \\ Y^Te & Y^TX \end{bmatrix},$$

yielding,

$$y^Te = 1 \quad \text{and} \quad Y^Te = 0. \tag{7.5}$$

Thus,

$$\alpha Q^{-1}SQ = \begin{bmatrix} \alpha y^TSe & \alpha y^TSX \\ \alpha Y^TSe & \alpha Y^TSX \end{bmatrix} = \begin{bmatrix} \alpha y^Te & \alpha y^TSX \\ \alpha Y^Te & \alpha Y^TSX \end{bmatrix} = \begin{bmatrix} \alpha & \alpha y^TSX \\ 0 & \alpha Y^TSX \end{bmatrix}, \tag{7.6}$$

where we have used $Se = e$. Since $Q^{-1}(\alpha S)Q$ is a similarity transformation of αS, then its eigenvalues are the same as αS and they are $\{\alpha, \alpha\lambda_2, ..., \alpha\lambda_n\}$.

In (7.6) above, the first eigenvalue is shown and since the matrix in (7.6) is upper triangular, the other eigenvalues are eigenvalues of $\alpha Y^T S X$.

Going back to (7.4) and computing the second term, we arrive at,

$$(1 - \alpha)Q^{-1}ev^T Q = (1 - \alpha) \begin{bmatrix} y^T \\ Y^T \end{bmatrix} ev^T \begin{bmatrix} e & X \end{bmatrix}$$

$$= \begin{bmatrix} (1-\alpha)y^T ev^T e & (1-\alpha)y^T ev^T X \\ (1-\alpha)Y^T ev^T e & (1-\alpha)Y^T ev^T X \end{bmatrix} = \begin{bmatrix} 1-\alpha & (1-\alpha)v^T X \\ 0 & 0 \end{bmatrix},$$

where we used (7.5) and the fact that v is a probability vector, giving $v^T e = 1$. Therefore, we arrive at the similarity transformation,

$$Q^{-1}G(\alpha)Q = \begin{bmatrix} \alpha & \alpha y^T SX \\ 0 & \alpha Y^T SX \end{bmatrix} + \begin{bmatrix} 1-\alpha & (1-\alpha)v^T X \\ 0 & 0 \end{bmatrix}$$

$$= \begin{bmatrix} 1 & \alpha y^T SX + (1-\alpha)v^T X \\ 0 & \alpha Y^T SX \end{bmatrix},$$

and as we found in (7.6), except for the first eigenvalue, the remaining eigenvalues of $\alpha Y^T SX$ are the same as those of αS. Thus, by the property of similarity transformation, the eigenvalues of $G(\alpha)$ are $\{1, \alpha\lambda_2, ..., \alpha\lambda_n\}$. □

In [13], J. Ding and A. Zhou find the eigenvalues of the Google matrix by a different method, considering the following equation for Google matrix,

$$G(\alpha) = \alpha S + (1 - \alpha)ue^T, \tag{7.7}$$

which is similar to (7.2) with the probability vector u being the personalization vector. Note that here ue^T produces a column stochastic matrix and for G to be a transition matrix, we assume S to also be a column stochastic matrix.

Theorem 7.3 (Corollary 3.1 in [13]) *If the eigenvalues of S are $\{1, \lambda_2, ..., \lambda_n\}$, then the eigenvalues of the Google matrix, $G(\alpha)$ are $\{1, \alpha\lambda_2, \alpha\lambda_3, ..., \alpha\lambda_n\}$.*

Proof. The authors first prove the statement for a general case and then apply it to the Google matrix. Here we explain their proof by directly applying the reasoning to the Google matrix. By using the fact that for any matrices, $A_1, A_2, ..., A_n$,

$$\det(A_1 A_2 ... A_n) = \det(A_1)\det(A_2)...\det(A_n),$$

we take the determinant of both sides of

$$\begin{bmatrix} I & 0 \\ v^T & 1 \end{bmatrix} \begin{bmatrix} I + uv^T & u \\ 0 & 1 \end{bmatrix} \begin{bmatrix} I & 0 \\ -v^T & 1 \end{bmatrix} = \begin{bmatrix} I & u \\ 0 & 1 + v^T u \end{bmatrix}, \tag{7.8}$$

for u and v being two n-dimensional column vectors, to obtain,

$$det(I + uv^T) = (1 + v^T u). \tag{7.9}$$

Replacing u with $A^{-1}u$ in (7.9), we obtain for an invertible matrix, $A : n \times n$,

$$det(A + uv^T) = det(A(I + A^{-1}uv^T)) = det(A)det(I + A^{-1}uv^T)$$
$$= det(A)(1 + v^T A^{-1} u). \tag{7.10}$$

Using $A := \lambda I - \alpha S$, $v := -(1 - \alpha)e$, and letting u be an eigenvector of A associated to eigenvalue α_1, we obtain by above,

$$det((\lambda I - \alpha S) - u(1 - \alpha)e^T) = (1 - (1 - \alpha)e^T (\lambda I - \alpha S)^{-1}u)det(\lambda I - \alpha S). \tag{7.11}$$

We then let λ be any scalar and not equal to any eigenvalue of αS. Observe that if the eigenvalues of S are $\{1, \lambda_2, \ldots, \lambda_n\}$, then the eigenvalues of αS are $\{\alpha, \alpha\lambda_2, \ldots, \alpha\lambda_n\}$, since α is an eigenvalue of A associated with eigenvalue u, then $\frac{1}{\lambda - \alpha}$, is an eigenvalue of A^{-1} corresponding to u. That is,

$$(\lambda I - \alpha S)^{-1}u = \frac{1}{\lambda - \alpha}u. \tag{7.12}$$

Then with (7.11), we have,

$$det\left(\lambda I - \left(\alpha S + (1 - \alpha)ue^T\right)\right)$$
$$= det\left((\lambda I - \alpha S) - (1 - \alpha)ue^T\right)$$
$$= \left(1 - (1 - \alpha)e^T(\lambda I - \alpha S)^{-1}u\right)det(\lambda I - \alpha S)$$
$$= \left(1 - (1 - \alpha)\frac{e^T}{\lambda - \alpha}u\right)(\lambda - \alpha)(\lambda - \alpha\lambda_2)\ldots(\lambda - \alpha\lambda_n)$$
$$= \frac{(\lambda - \alpha) - (1 - \alpha)e^T u)(\lambda - \alpha)(\lambda - \alpha\lambda_2)\ldots(\lambda - \alpha\lambda_n)}{\lambda - \alpha}$$
$$= \left((\lambda - \alpha) - (1 - \alpha)e^T u\right)(\lambda - \alpha\lambda_2)\ldots(\lambda - \alpha\lambda_n).$$

Hence, setting the above equal to zero we have, $\{\alpha + (1 - \alpha)e^T u, \alpha\lambda_2, \ldots, \alpha\lambda_n\}$ for eigenvalues of $\alpha S + (1 - \alpha)ue^T$. Note that u being a probability vector, implies that $e^T u = 1$. Thus, the first eigenvalue is 1. $\qquad\square$

In [15], T. Haveliwala, S. Kamvar study properties of the second eigenvalue of the Google matrix. In Section 7.3 we discuss how their results may be used to detect if a website is link spamming. Recall that link spamming refers to web developers trying to increase their PageRank by creating many inlinks to their site, since based on the PageRank formula (7.1), that would improve their PageRank.

In [15], the authors give the Google matrix (7.2) as,

$$A = (cP + (1 - c)E)^T, \tag{7.13}$$

where $E: = ev^T$ and $P: n \times n$ are row stochastic matrices. Observe that (7.13) is equivalent to (7.2), by letting $c: = \alpha$ and $P^T: = S$ and noting that P^T as well as ve^T are column stochastic matrices. Based on this notation, they achieve the following results.

Theorem 7.4 *(Theorem 1 in [15]) If λ_2 is the second eigenvalue of the Google matrix with damping factor, c, then $|\lambda_2| \leq c$.*

Proof. Since $c \in [0, 1]$, we may consider three cases for c: $c = 0, c = 1$, and $0 < c < 1$. If $c = 0$, matrix A becomes,

$$E^T = ve^T = \begin{bmatrix} v_1 & v_2 & \cdots & v_n \end{bmatrix} \begin{bmatrix} 1 \\ 1 \\ \cdot \\ \cdot \\ 1 \end{bmatrix} = \begin{bmatrix} v_1 + v_2 + \ldots + v_n \end{bmatrix}.$$

Therefore, matrix A may only have one eigenvalue, which is λ_1 and $\lambda_2 = 0$. For the case $c = 1$, since for the Google matrix, the dominant eigenvalue is $\lambda_1 = 1$, then $|\lambda_2| \leq |\lambda_1| = 1$, implying $|\lambda_2| \leq c$.

To prove the statement for $0 < c < 1$, we note that here the authors assume that the second eigenvalue of the Google matrix satisfies $|\lambda_2| \leq \lambda_1$, instead of $|\lambda_2| < \lambda_1$ and thus they first prove $|\lambda_2| < 1$. They apply a theorem that states if a Markov chain, M, has only one irreducible, closed, aperiodic subchain, S, then M has a unique eigenvector with eigenvalue 1. Details are given as follows. Since for any matrix, A, eigenvalues of A and A^T are equivalent, they verify $|\lambda_2| < 1$ for the transition matrix, A^T.

Let U be the set of states that have nonzero components in the personalization vector, v. Denote S to be the set of states in the Markov chain that can be reached from a state in U. The authors prove that S is the needed unique irreducible, closed, aperiodic subchain of the Markov chain.

By its construction, the set S is closed and it is also irreducible since by the definition of U, every state in the Markov chain reaches a state in U, which prevents any subset of S to be closed and hence all states in S communicate with each other. It is known (see for example, Pg. 82 in [16]) that in an irreducible closed subset of a Markov chain, any two distinct states of the same class have the same period. Every state in the Markov chain reaches a state in U. This includes states in U itself; thus, every $u \in U$ has a loop making its period 1 and since U contains S, then S is aperiodic. Moreover, S must be the only irreducible, closed, aperiodic subset of the Markov chain, since there is only one set U and U contains every closed subset in the Markov chain. In addition, every closed subset that contains U has to contain S, implying that S has to be unique.

Therefore, we obtain that $|\lambda_2| < 1$ and the dominant eigenvalue of A being $\lambda_1 = 1$, implies $|\lambda_2| < |\lambda_1|$. Let x_2 be the eigenvector of A corresponding to λ_2. Then,

$$Ax_2 = \lambda_2 x_2 \quad \text{giving} \quad cP^T x_2 + (1-c)E^T x_2 = \lambda_2 x_2. \qquad (7.14)$$

Notice that for any matrix B, if x_i is an eigenvector corresponding to λ_i and y_i is an eigenvalue of B^T corresponding to λ_j, where $\lambda_i \neq \lambda_j$, then $x_i^T y_j = 0$. This may be directly reasoned as follows,

$$Ax_i = \lambda_i x_i \quad \Rightarrow \quad x_i^T A^T = \lambda_i x_i^T \quad \Rightarrow \quad x_i^T \left(A^T y_i \right) = \lambda_i x_i^T y_i$$
$$\Rightarrow \quad x_i^T \left(\lambda_j y_i \right) = \lambda_i x_i^T y_i$$
$$\Rightarrow \quad x_i^T y_i (\lambda_j - \lambda_i) = 0 \quad \Rightarrow \quad x_i^T y_i = 0.$$

Since A^T is a row stochastic matrix, then $A^T e = e$ and its first eigenvalue being 1 implies $x_1 = e$. Hence, noting that $\lambda_2 \neq \lambda_1$, we apply the fact above to A^T to obtain, $e^T x_2 = 0$. Therefore, (7.14) becomes,

$$cP^T x_2 = \lambda_2 x_2,$$

implying that x_2 is an eigenvector of P^T with eigenvalue $\frac{\lambda_2}{c}$. Since P is a stochastic matrix, then the modulus of eigenvalues of P^T must be less than or equal to 1. Thus, $\left| \frac{\lambda_2}{c} \right| \leq 1$, yielding the result, $|\lambda_2| \leq c$. $\qquad \square$

For the next theorem we need the following result proved on Pg. 126 of [18].

Theorem 7.5 *The multiplicity of eigenvalue 1 for the transition matrix of a finite Markov chain is equal to the number of the chain's irreducible, closed subsets.*

Theorem 7.6 *(Theorem 2 in [15]) The modulus of the second eigenvalue of A is equal to c if P has at least two irreducible closed subsets.*

Proof. Suppose that P^T has at least two irreducible closed subsets. If $c = 0$, then by the reasoning in the proof of Theorem 7.4, $\lambda_2 = c$. For $c = 1$, we use Theorem 7.5 to conclude that $\lambda_1 = 1$ has multiplicity 2 and $|\lambda_2| = 1 = c$. In the case $0 < c < 1$, we prove that $|\lambda_2| \geq c$ so that with Theorem 7.4, we obtain, $|\lambda_2| = c$.

Since eigenvalue 1 of P^T has multiplicity 2, then there exists two linearly independent eigenvectors of P^T for eigenvalue 1. We denote these by y_1 and y_2 and use them to form an eigenvector, x_i of P^T that is orthogonal to e. Let,

$$k_1 = y_1^T e, \quad \text{and} \quad k_2 = y_2^T e.$$

In the case $k_1 = 0$ or $k_2 = 0$, letting $x_i = y_1$ or $x_i = y_2$, respectively would yield the orthogonality of x_i with e. If $k_1, k_2 > 0$, then we let $x_i = \frac{y_1}{k_1} - \frac{y_2}{k_2}$ to obtain,

$$x_i^T e = \left[\frac{y_1^T}{k_1} - \frac{y_2^T}{k_2}\right] e = \frac{y_1^T e}{k_1} - \frac{y_2^T e}{k_2} = 1 - 1 = 0.$$

Thus, eigenvector x_i, by the above construction, is orthogonal to e. Then, x_i is an eigenvector of A, since,

$$Ax_i = cP^T x_i + (1-c)E^T x_i = cP^T x_i + (1-c)ve^T x_i = cP^T x_i = c\gamma_i x_i,$$

where, γ_i is the eigenvalue of P^T corresponding to x_i. Because, $x_i = y_1$ or $x_i = y_2$, then $\gamma_i = 1$ implying that c is the eigenvalue of A corresponding to x_i. Hence, $\lambda_i = c$. For a column stochastic matrix such as A, $|\lambda_1| \geq |\lambda_2| \geq \ldots \geq |\lambda_n|$. Thus, $|\lambda_2| \geq |\lambda_i| = c$ and we obtain the result. □

7.2 Computing the PageRank

We now concentrate on the power method and explain results on determining the PageRank. The power method is an iterative method given by,

$$\pi_k^T = \pi_{k-1}^T G, \tag{7.15}$$

where G is the Google matrix and vector, π_{k-1}^T is a probability vector, implying, $\pi_{k-1}^T e = 1$. In [22], A. Langville and C. Meyer use $S := H + av^T$ in (7.2), giving,

$$G(\alpha) = \alpha H + (\alpha a + (1-\alpha)e)v^T, \tag{7.16}$$

where, a is a vector with $a_i = 1$ if the ith row of H corresponds to a dangling node and $a_i = 0$ if the ith page has outlinks. Recall that H stands for the hyperlink matrix.

Theorem 7.7 (Theorem 2.1 in [22]) *Let $x^T := v^T(I - \alpha H)^{-1}$, then $\pi^T = \frac{x^T}{x^T e}$ is the PageRank vector for the Google matrix.*

Proof. Since the vector x is nonnegative, it is left to show that $\pi^T = \frac{x^T}{x^T e}$ satisfies $\pi^T G = \pi^T$ and $\pi^T e = 1$ to prove that it is the PageRank of G. By the form of π^T given in the theorem, $\pi^T e = 1$ follows immediately. We may write $\pi^T G = \pi^T$ as $0^T = \pi^T(I - G)$, which by our choice of π^T becomes, $0^T = x^T(I - G)$. Using (7.16) for G and by the given form of x^T we have,

$$
\begin{aligned}
x^T(I - G) &= v^T(I - \alpha H)^{-1}(I - G) \\
&= v^T(I - \alpha H)^{-1}(I - \alpha H) - v^T(I - \alpha H)^{-1}(\alpha a + (1 - \alpha)e)v^T \\
&= v^T - v^T(I - \alpha H)^{-1}(\alpha(a - e) + e)v^T \\
&= v^T - v^T(I - \alpha H)^{-1}(-\alpha He + e)v^T \\
&= v^T - v^T(I - \alpha H)^{-1}(I - \alpha H)ev^T \\
&= v^T - v^T ev^T = v^T - v^T = 0^T,
\end{aligned}
$$

where we have applied the fact that v is a probability vector giving $v^T e = 1$ and also we obtain $e - a = He$ based on the structure of vector a corresponding to dangling and nondangling nodes of the system, that is, noting that the row in H corresponding to a dangling node is a zero row.   □

Now we turn our attention to dangling nodes and find how the computation of the PageRank may be expedited if these nodes are lumped together into a single node and by permutation, the zero rows are placed at the bottom of the matrix. That is, we partition the hyperlink matrix, H as,

$$
H = \begin{array}{c} \\ \text{ND} \\ \text{D} \end{array} \overset{\displaystyle \begin{array}{cc} \text{ND} & \text{D} \end{array}}{\left[\begin{array}{cc} H_{11} & H_{12} \\ 0 & 0 \end{array} \right]}.
$$

where, ND and D stand for nondangling and dangling nodes, respectively. Note that there are no links from dangling nodes to nondangling nodes or to dangling nodes, and thus we have all zeros for the bottom two blocks. Hence, H_{11} and H_{12} are $k \times k$ matrices representing the k nondangling nodes and there are $n - k$ dangling nodes. The nonzero rows in the hyperlink matrix add to 1 and since the matrix S in (7.2) is required to be a row stochastic matrix, Ipsen and Selee [17], replace every zero row by the nonzero vector, $w^T = [w_1^T \quad w_2^T]$, where $w_1 : k \times 1$ and $w_2 : (n-k) \times 1$ and $\|w\|_1 = 1$. Then with $d^T = [0^T \quad e^T]$,

$$S = H + dw^T = \begin{bmatrix} H_{11} & H_{12} \\ 0 & 0 \end{bmatrix} + \begin{bmatrix} 0 \\ e \end{bmatrix} [w_1^T \quad w_2^T]$$

$$= \begin{bmatrix} H_{11} & H_{12} \\ 0 & 0 \end{bmatrix} + \begin{bmatrix} 0 & 0 \\ ew_1^T & ew_2^T \end{bmatrix} = \begin{bmatrix} H_{11} & H_{12} \\ ew_1^T & ew_2^T \end{bmatrix},$$

making S a row stochastic matrix since,

$$ew_1^T + ew_2^T = \begin{bmatrix} 1 \\ \cdot \\ \cdot \\ 1 \end{bmatrix} [w_{11} \quad w_{12} \quad \ldots \quad w_{1k}] + \begin{bmatrix} 1 \\ \cdot \\ \cdot \\ 1 \end{bmatrix} [w_{2,(n-k)} \quad w_{2,(n-k-1)} \quad \ldots \quad w_{2,n}]$$

$$= \begin{bmatrix} w_{11} + w_{12} + \ldots + w_{1k} + w_{2,(n-k)} + \ldots + w_{2,n} \\ \cdot \\ \cdot \\ w_{n1} + w_{n2} + \ldots + w_{nk} + w_{n,(n-k)} + \ldots + w_{n,n} \end{bmatrix} = \begin{bmatrix} 1 \\ \cdot \\ \cdot \\ 1 \end{bmatrix} = e.$$

The Google matrix then becomes,

$$G = \alpha S + (1-\alpha)ev^T = \begin{bmatrix} \alpha H_{11} & \alpha H_{12} \\ \alpha ew_1^T & \alpha ew_2^T \end{bmatrix} + (1-\alpha) \begin{bmatrix} 1 \\ \cdot \\ \cdot \\ 1 \end{bmatrix} [v_1 \quad \ldots \quad v_n]$$

$$= \begin{bmatrix} \alpha H_{11} + (1-\alpha)(v_1 + \ldots + v_k) & \alpha H_{12} + (1-\alpha)(v_{n-k} + \ldots + v_n) \\ \alpha ew_1^T + (1-\alpha)(v_1 + \ldots + v_k) & \alpha ew_2^T + (1-\alpha)(v_{n-k} + \ldots + v_n) \end{bmatrix}$$

$$= \begin{bmatrix} G_{11} & G_{12} \\ eu_1^T & eu_2^T \end{bmatrix}, \tag{7.17}$$

where,

$$u^T = [u_1^T \quad u_2^T] = \alpha w^T + (1-\alpha)v^T = \alpha [w_1^T \quad w_2^T] + (1-\alpha)[v_1^T \quad v_2^T].$$

In [17], Ipsen and Selee use the form of hyperlink and Google matrix given in (7.17) and compute the PageRank first given as one vector and then partitioned as $\pi^T = [\pi_1^T \quad \pi_2^T]$, with π_1 and π_2 being the PageRanks of the nondangling and dangling nodes, respectively. Before we explain their results, we need the following equation,

$$(A + uv^H)^{-1} = A^{-1} - \frac{A^{-1}uv^H A^{-1}}{1 + v^H A^{-1} u}, \tag{7.18}$$

referred to as the Sherman-Morison formula, which can be used on a non-singular matrix, A and nonzero number, $1 + v^H A^{-1} u$, with v^H denoting the Hermitian of v, meaning conjugate transpose of v. One can verify that the expression on the right side of the above formula is indeed the inverse of $A + uv^H$ by multiplying (7.18) from the left by $A + uv^H$ and noting that $v^H A^{-1} u = uv^H A^{-1}$ to obtain the identity matrix and following the same steps when multiplying (7.18) by $A + uv^H$ from the right.

Recall that e is a column vector having all entries be 1. The column vector with all entries being zero except for the ith entry being 1, is denoted as e_i. For a vector, σ, to be a stationary distribution of a row stochastic matrix, A, need $\sigma A = \sigma$ and $\|\sigma\|_1 = 1$.

Theorem 7.8 (Theorem 3.2 in [17]) *Let σ^T be the stationary distribution of Google matrix, G, then using the partition, $\sigma^T = [\sigma_{1:k}^T \quad \sigma_{k+1}]$, where σ_{k+1} is a scalar, the PageRank vector is given by,*

$$\pi^T = \begin{bmatrix} \sigma_{1:k}^T & \sigma^T \begin{bmatrix} G_{12} \\ u_2^T \end{bmatrix} \end{bmatrix}. \tag{7.19}$$

Proof. Let

$$\begin{bmatrix} I_k & 0 \\ 0 & L \end{bmatrix},$$

where,

$$L = I_{n-k} - \frac{1}{n-k} \hat{e} e^T,$$

with $\hat{e} = e - e_1$. We use X for the matrix in the similarity transformation of G and find,

$$XGX^{-1} = \begin{bmatrix} I_k & 0 \\ 0 & L \end{bmatrix} \begin{bmatrix} G_{11} & G_{12} \\ eu_1^T & eu_2^T \end{bmatrix} \begin{bmatrix} I_k & 0 \\ 0 & L^{-1} \end{bmatrix} = \begin{bmatrix} G_{11} & G_{12}L^{-1} \\ Leu_1^T & Leu_2^T L^{-1} \end{bmatrix}.$$

We may apply the Sherman-Morison formula (7.18) to L to obtain,

$$L^{-1} = \left(I_{n-k} + \left(-\frac{1}{n-k} \hat{e} e^T \right) \right)^{-1} = I - \frac{I\hat{e} e^T \left(\frac{-1}{n-k} \right)}{1 + e^T \hat{e} \left(\frac{-1}{n-k} \right)}. \tag{7.20}$$

Note that,

$$e^T \hat{e} = \begin{bmatrix} 1 & 1 & .. & 1 \end{bmatrix} \begin{bmatrix} 0 \\ 1 \\ . \\ . \\ 1 \end{bmatrix} = 0 + 1 + ... + 1 = n - k - 1,$$

giving,

$$L^{-1} = I - \frac{\hat{e}e^T \left(\frac{-1}{n-k}\right)}{1 + (n-k)\left(\frac{-1}{n-k}\right) + \frac{1}{n-k}} = I + \hat{e}e^T.$$

By the same reasoning as above, $e^T e = n - k$ and by definition, $e - \hat{e} = e_1$. Thus,

$$XGX^{-1} = \begin{bmatrix} G_{11} & G_{12}(I_{n-k} + \hat{e}e^T) \\ eu_1^T - \frac{1}{n-k}\hat{e}e^T eu_1^T & \left(eu_2^T - \frac{1}{n-k}\hat{e}e^T eu_2^T\right)(I_{n-k} + \hat{e}e^T) \end{bmatrix}$$

$$= \begin{bmatrix} G_{11} & G_{12}(I_{n-k} + \hat{e}e^T) \\ eu_1^T - \hat{e}u_1^T & \left(eu_2^T - \hat{e}u_2^T\right)(I_{n-k} + \hat{e}e^T) \end{bmatrix}$$

$$= \begin{bmatrix} G_{11} & G_{12}\left(I_{n-k} + \hat{e}e^T\right) \\ e_1 u_1^T & e_1 u_2^T \left(I_{n-k} + \hat{e}e^T\right) \end{bmatrix}.$$

For better presentation, we denote I_{n-k} as I and notice that,

$$G_{12}\left(I_{n-k} + \hat{e}e^T\right) e_1 = G_{12}e_1 + G_{12}\hat{e}e^T e_1 = G_{12}\left(e_1 + \hat{e}\right)e^T e_1 = G_{12}e,$$

where it can be seen that $e^T e_1 = 1$. Similarly,

$$u_2^T\left(I + \hat{e}e^T\right) e_1 = u_2^T e_1 + u_2^T \hat{e}e^T e_1 = u_2^T (e_1 + \hat{e}) = u_2^T e.$$

Hence, since G_{11} contains the first k entries of G,

$$XGX^1 = \begin{bmatrix} G_{11} & G_{12}\left(I + \hat{e}e^T\right) e_1 & G_{12}\left(I + \hat{e}e^T\right) e_2 & ... & G_{12}\left(I + \hat{e}e^T\right) e_{n-k} \\ e_1 u_1^T & e_1 u_2^T \left(I + \hat{e}e^T\right) e_1 & e_1 u_2^T \left(I + \hat{e}e^T\right) e_2 & ... & e_1 u_2^T \left(I + \hat{e}e^T\right) e_{n-k} \end{bmatrix}$$

$$= \begin{bmatrix} G_{11} & G_{12}e & G_{12}\left(I + \hat{e}e^T\right) e_2 & ... & G_{12}\left(I + \hat{e}e^T\right) e_{n-k} \\ e_1 u_1^T & u_2^T e & e_1 u_2^T \left(I + \hat{e}e^T\right) e_2 & ... & e_1 u_2^T \left(I + \hat{e}e^T\right) e_{n-k} \end{bmatrix}.$$

Denoting,

$$
G^{(1)} = \begin{bmatrix} G_{11} & G_{12}e \\ u_1^T & u_2^T e \end{bmatrix},
$$

we have,

$$
XGX^{-1} = \begin{bmatrix} G^{(1)} & G^{(2)} \\ 0 & 0 \end{bmatrix}, \tag{7.21}
$$

where

$$
G^{(2)} = \begin{bmatrix} G_{12} \left(I + \hat{e}e^T \right) \\ u_2^T \left(I + \hat{e}e^T \right) \end{bmatrix} \begin{bmatrix} e_2 & \cdots & e_{n-k} \end{bmatrix}.
$$

Based on the structure in (7.21) and nonzero rows, we can see that $G^{(1)}$ has the same eigenvalues as XGX^{-1} and thus based on similarity transformation property, $G^{(1)}$ has the same eigenvalues as G. Since $G^{(1)}$ has $k+1$ nonzero rows, then G has at most $k+1$ nonzero eigenvalues. The vector σ^T being the stationary distribution of $G^{(1)}$, yields,

$$
\begin{bmatrix} \sigma^T & \sigma^T G^{(2)} \end{bmatrix} XGX^{-1} = \begin{bmatrix} \sigma^T & \sigma^T G^{(2)} \end{bmatrix} \begin{bmatrix} G^{(1)} & G^{(2)} \\ 0 & 0 \end{bmatrix} = \begin{bmatrix} \sigma^T G^{(1)} & \sigma^T G^{(2)} \end{bmatrix}
$$
$$
= \begin{bmatrix} G^{(1)} & \sigma^T G^{(2)} \end{bmatrix}, \tag{7.22}
$$

implying that $\begin{bmatrix} \sigma^T & \sigma^T G^{(2)} \end{bmatrix}$ is a left eigenvector of XGX^{-1} corresponding to the eigenvalue 1. Also multiplying both sides of (7.22) by X makes, $\begin{bmatrix} \sigma^T & \sigma^T G^{(2)} \end{bmatrix} X$ an eigenvector of G corresponding to eigenvalue 1. We let,

$$
\hat{\pi}^T = \begin{bmatrix} \sigma^T & \sigma^T G^{(2)} \end{bmatrix} X, \tag{7.23}
$$

denote the above eigenvector and prove that $\hat{\pi}^T$ is the PageRank. Recall that for this purpose, we need to confirm that $\hat{\pi}^T \geq 0$, $\|\hat{\pi}\|_1 = 1$ and $\hat{\pi}^T G = \hat{\pi}^T$. The first condition follows by the assumption that $\sigma^T \geq 0$. To verify the other two conditions, we proceed by simplifying the expressions in $\hat{\pi}^T$. Note that since σ^T satisfies, $\sigma^T G^{(1)} = \sigma^T$, then σ is a column with $k+1$ entries and the top left block of X has k rows. Hence, we write,

$$
\hat{\pi}^T = \begin{bmatrix} \sigma_{1:k}^T & \sigma_{k+1} & \sigma^T G^{(2)} \end{bmatrix} \begin{bmatrix} I_k & 0 \\ 0 & L \end{bmatrix} = \begin{bmatrix} \sigma_{1:k}^T & \left(\sigma_{k+1} & \sigma^T G^{(2)} \right) L \end{bmatrix}. \tag{7.24}
$$

To match the matrix L with the setup of $\begin{pmatrix} \sigma_{k+1} & \sigma^T G^{(2)} \end{pmatrix}$, we write,

$$L = I - \frac{1}{n-k}\hat{e}\hat{e}^T = I - \frac{1}{n-k}\begin{bmatrix} 0 \\ 1 \\ 1 \\ \vdots \\ 1 \end{bmatrix}\begin{bmatrix} 1 & \cdots & 1 \end{bmatrix} = I - \frac{1}{n-k}\begin{bmatrix} 0 & 0 & \cdots & 0 \\ 1 & 1 & \cdots & 1 \\ \vdots & \vdots & \ddots & \vdots \\ 1 & \cdot & \cdot & 1 \end{bmatrix}$$

$$= \begin{bmatrix} 1 & 0 & \cdots & 0 \\ \frac{-1}{n-k}e & I - \frac{1}{n-k}ee^T & & \end{bmatrix},$$

where e has $n-k-1$ rows and ee^T is an $(n-k-1) \times (n-k-1)$ matrix having 1 for each of its entries. Therefore,

$$\begin{pmatrix} \sigma_{k+1} & \sigma^T G^{(2)} \end{pmatrix} L = \begin{bmatrix} \sigma_{k+1} & \sigma^T G^{(2)} \end{bmatrix}\begin{bmatrix} 1 & 0 \\ \frac{-1}{n-k}e & I - \frac{1}{n-k}ee^T \end{bmatrix}$$

$$= \begin{bmatrix} \sigma_{k+1} - \frac{1}{n-k}\sigma^T G^{(2)}e & \sigma^T G^{(2)}\left(I - \frac{1}{n-k}ee^T\right) \end{bmatrix}.$$

Recall the expression for $G^{(2)}$ and note that,

$$\begin{bmatrix} e_2 & \cdots & e_{n-k} \end{bmatrix}e = \begin{bmatrix} e_2 & \cdots & e_{n-k} \end{bmatrix}\begin{bmatrix} 1 \\ 1 \\ \vdots \\ 1 \end{bmatrix}$$

$$= e_2 + \ldots + e_{n-k} = \begin{bmatrix} 0 \\ 1 \\ 0 \\ \vdots \\ \vdots \\ 0 \end{bmatrix} + \begin{bmatrix} 0 \\ 0 \\ 1 \\ \vdots \\ \vdots \\ 0 \end{bmatrix} + \ldots + \begin{bmatrix} 0 \\ 0 \\ \vdots \\ 1 \\ 0 \\ 1 \end{bmatrix} = \begin{bmatrix} 0 \\ 1 \\ 1 \\ \vdots \\ \vdots \\ 1 \end{bmatrix} = \hat{e},$$

and since $\hat{e}\hat{e}^T$ is an $(n-k) \times (n-k)$ matrix having zeros for all entries on the first row and 1 for all entries in every other row, we have,

$$\left(I + \hat{e}\hat{e}^T\right)\hat{e} = \begin{bmatrix} 1 & 0 & \cdot & \cdot & \cdot & 0 \\ 1 & 2 & 1 & \cdot & \cdot & 1 \\ 1 & 1 & 2 & 1 & \cdot & 1 \\ \cdot & \cdot & \cdot & \cdot & \cdot & \cdot \\ 1 & \cdot & \cdot & \cdot & \cdot & 2 \end{bmatrix}\begin{bmatrix} 0 \\ 1 \\ \vdots \\ \vdots \\ 1 \end{bmatrix} = \begin{bmatrix} 0 \\ 2+1+\ldots+1 \\ \cdots \\ 2+1+\ldots+1 \end{bmatrix} = \begin{bmatrix} 0 \\ n-k \\ \cdot \\ n-k \end{bmatrix}$$

$$= (n-k)\hat{e},$$

leading to,

$$G^{(2)}e = \begin{bmatrix} G_{12} \\ u_2^T \end{bmatrix}(n-k)\hat{e} = \begin{bmatrix} G_{12} \\ u_2^T \end{bmatrix}(n-k)(e-e_1). \tag{7.25}$$

We use (7.25) to further find,

$$\frac{1}{n-k}\sigma^T G^{(2)}e = \sigma^T \begin{bmatrix} G_{12} \\ u_2^T \end{bmatrix}e - \sigma^T \begin{bmatrix} G_{12} \\ u_2^T \end{bmatrix}e_1.$$

Since,

$$\sigma^T \begin{bmatrix} G_{12} \\ u_2^T \end{bmatrix}e = \sigma_{k+1},$$

we obtain, using (7.25) again,

$$\sigma_{k+1} = \frac{1}{n-k}\sigma^T G^{(2)}e + \sigma^T \begin{bmatrix} G_{12} \\ u_2^T \end{bmatrix}e_1$$

$$= \sigma^T \begin{bmatrix} G_{12} \\ u_2^T \end{bmatrix}(e-e_1) + \sigma^T \begin{bmatrix} G_{12} \\ u_2^T \end{bmatrix}e_1 = \sigma^T \begin{bmatrix} G_{12} \\ u_2^T \end{bmatrix}e.$$

Hence, we have,

$$\hat{\pi} = \begin{bmatrix} \sigma_{1:k}^T & \sigma^T \begin{bmatrix} G_{12} \\ u_2^T \end{bmatrix} \end{bmatrix},$$

and by the uniqueness of PageRank we obtain that $\hat{\pi}$ is the PageRank of G given in (7.19). $\qquad\square$

For the proof of the theorem below we introduce the concept of M-matrix. Let $\sigma(A)$ denote the spectral radius of matrix A, meaning the largest eigenvalue of A. If matrix $A : n \times n$ can be written in the form, $A = sI - B$, where B is a nonnegative matrix such that $\sigma(B) \leq s$, then A is called an M-matrix. If A is an M-matrix, then its inverse exists and is nonnegative (see Theorem 1 in [26]).

Theorem 7.9 (Theorem 4.1 in [17]) *The PageRank of the nondangling nodes and dangling nodes are,*

$$\pi_1^T = \left((1-\alpha)v_1^T + \rho w_1^T\right)(I - \alpha H_{11})^{-1}, \tag{7.26}$$

$$\pi_2^T = \alpha\pi_1^T H_{12} + (1-\alpha)v_2^T + \alpha\left(1 - \|\pi_1\|_1\right)w_2^T, \tag{7.27}$$

respectively, where,

$$\rho = \alpha \frac{1 - (1 - \alpha)v_1^T (I - \alpha H_{11})^{-1} e}{1 + \alpha w_1^T (I - \alpha H_{11})^{-1} e} \geq 0. \tag{7.28}$$

Proof. Since G is a row stochastic matrix, then $Ge = e$ and letting $S = H + dw^T$ in (7.2) we arrive at,

$$e = Ge = \alpha \left(H + dw^T \right) e + (1 - \alpha)ev^T e,$$

leading to,

$$\left(I - \alpha H - \alpha dw^T \right) e = (1 - \alpha)ev^T e.$$

Then multiplying both sides by e^{-1} and solving for e yield,

$$e = \left(\frac{I - \alpha H - \alpha dw^T}{1 - \alpha} \right) \left(v^T \right)^{-1},$$

implying,

$$e^{-1} = (1 - \alpha)v^T \left(I - \alpha H - \alpha dw^T \right)^{-1},$$

where we applied the facts that $(AB)^{-1} = B^{-1}A^{-1}$ and $(kA)^{-1} = \frac{1}{k}A^{-1}$. We now verify the conditions needed for π to be a PageRank. Condition, $\|\pi\|_1 = 1$ is equivalent to $\pi^T e = 1$, which holds if and only if,

$$\pi^T = e^{-1} = (1 - \alpha)v^T \left(I - \alpha H - \alpha dw^T \right)^{-1}. \tag{7.29}$$

Next we must verify that π is nonnegative. We first simplify the expression in (7.29) by denoting $R := I - \alpha H$ and applying the Sherman-Morison formula (7.18),

$$\pi^T = (1 - \alpha)v^T \left(R^{-1} - \frac{R^{-1}(-\alpha dw^T)R^{-1}}{1 + w^T R^{-1}(-\alpha d)} \right)$$

$$= (1 - \alpha)v^T R^{-1} + \frac{\alpha(1 - \alpha)v^T R^{-1}d}{1 - \alpha w^T R^{-1}d} w^T R^{-1}.$$

Observe that R is an M-matrix, since denoting the nonnegative matrix $B := \alpha H$ in the definition of M-matrix, we have $\sigma(B) = \alpha(1) = \alpha$ which is less than $s = 1$. Hence, $R^{-1} \geq 0$ and we obtain $w \geq 0$, $d \geq 0$ and $v \geq 0$. Therefore, the first term and the numerator of the second term are nonnegative and it is left to verify that $1 - \alpha w^T R^{-1} d > 0$. Since,

$$R = I - \alpha H = \begin{bmatrix} I_{k\times k} & 0 \\ 0 & I_{(n-k)\times(n-k)} \end{bmatrix} - \begin{bmatrix} \alpha H_{11} & \alpha H_{12} \\ 0 & 0 \end{bmatrix} = \begin{bmatrix} I - \alpha H_{11} & -\alpha H_{12} \\ 0 & I \end{bmatrix},$$

then,

$$R^{-1} = \begin{bmatrix} (I - \alpha H_{11})^{-1} & \alpha(I - \alpha H_{11})^{-1} H_{12} \\ 0 & I \end{bmatrix}, \tag{7.30}$$

and

$$1 - \alpha w^T R^{-1} d = 1 - \alpha \left(\begin{bmatrix} w_1^T & w_2^T \end{bmatrix} \begin{bmatrix} (I - \alpha H_{11})^{-1} & \alpha(I - \alpha H_{11})^{-1} H_{12} \\ 0 & I \end{bmatrix} \begin{bmatrix} 0 \\ e \end{bmatrix} \right)$$

$$= 1 - \alpha \left(\begin{bmatrix} w_1^T(I - \alpha H_{11})^{-1} & w_1^T\alpha(I - \alpha H_{11})^{-1} H_{12} + w_2^T \end{bmatrix} \begin{bmatrix} 0 \\ e \end{bmatrix} \right)$$

$$= 1 - \alpha \left(w_1^T\alpha(I - \alpha H_{11})^{-1} H_{12}e + w_2^T e \right).$$

Notice that $H_{11}e + H_{12}e = e$ gives,

$$\alpha(I - \alpha H_{11})^{-1} H_{12}e = e - (I - \alpha H_{11})^{-1}(1 - \alpha)e, \tag{7.31}$$

and since $w_1^T e + w_2^T e = 1$, we obtain,

$$1 - \alpha w^T R^{-1} d = 1 - \alpha w_1^T e + \alpha(1 - \alpha)w_1^T(I - \alpha H_{11})^{-1}e - \alpha w_2^T e$$

$$= 1 - \alpha + \alpha(1 - \alpha)w_1^T(I - \alpha H_{11})^{-1}e$$

$$= (1 - \alpha)\left(1 + \alpha w_1^T(I - \alpha H_{11})^{-1}e\right).$$

By $I - \alpha H_{11}$ being an M-matrix we achieve, $1 - \alpha w^T R^{-1} d \geq 0$. Then the expression in (7.31) and the fact that v is a probability vector are used to rewrite the numerator of the second term in the expression for π^T as follows,

$$v^T R^{-1} d = \begin{bmatrix} v_1^T & v_2^T \end{bmatrix} \begin{bmatrix} (I - \alpha H_{11})^{-1} & \alpha(I - \alpha H_{11})^{-1} H_{12} \\ 0 & I \end{bmatrix} \begin{bmatrix} 0 \\ e \end{bmatrix}$$

$$= \begin{bmatrix} v_1^T(I - \alpha H_{11})^{-1} & v_1^T\alpha(I - \alpha H_{11})^{-1} H_{12} + v_2^T \end{bmatrix} \begin{bmatrix} 0 \\ e \end{bmatrix}$$

$$= v_1^T\alpha(I - \alpha H_{11})^{-1} H_{12}e + v_2^T e = v_1^T e - v_1^T(1 - \alpha)(I - \alpha H_{11})^{-1}e + v_2^T e$$

$$= 1 - v_1^T(1 - \alpha)(I - \alpha H_{11})^{-1}e.$$

Since $v^T R^{-1} d \geq 0$, then the above expression is nonnegative and we may write,

$$\pi^T = (1 - \alpha)v^T R^{-1} + \alpha \frac{1 - (1 - \alpha)v_1^T(I - \alpha H_{11})^{-1}e}{1 + \alpha w_1^T(I - \alpha H_{11})^{-1}e} w_1^T R^{-1}.$$

Next letting ρ be defined as in (7.28), we have,

$$\pi^T = \left((1 - \alpha)v^T + \rho w_1^T\right) R^{-1},$$

and by the first components of v and R^{-1} we obtain (7.26). Now for π_2 from (7.29) we obtain,

$$\pi^T \left(I - \alpha H - \alpha d w^T\right) = (1 - \alpha)v^T, \tag{7.32}$$

where the left hand side may be expanded as follows,

$$\begin{bmatrix} \pi_1^T & \pi_2^T \end{bmatrix} \left(\begin{bmatrix} I_{k \times k} & 0 \\ 0 & I_{(n-k) \times (n-k)} \end{bmatrix} - \begin{bmatrix} \alpha H_{11} & \alpha H_{12} \\ 0 & 0 \end{bmatrix} - \alpha \begin{bmatrix} 0 \\ e \end{bmatrix} \begin{bmatrix} w_1^T & w_2^T \end{bmatrix} \right)$$

$$= \begin{bmatrix} \pi_1^T & \pi_2^T \end{bmatrix} \left(\begin{bmatrix} I - \alpha H_{11} & -\alpha H_{12} \\ 0 & I \end{bmatrix} - \begin{bmatrix} 0 & 0 \\ \alpha e w_1^T & \alpha e w_2^T \end{bmatrix} \right)$$

$$= \begin{bmatrix} \pi_1^T & \pi_2^T \end{bmatrix} \begin{bmatrix} I - \alpha H_{11} & -\alpha H_{12} \\ -\alpha e w_1^T & I - \alpha e w_2^T \end{bmatrix}$$

$$= \begin{bmatrix} \pi_1^T - \pi_1^T \alpha H_{11} - \pi_2^T \alpha e w_1^T & -\alpha \pi_1^T H_{12} + \pi_2^T - \alpha \pi_2^T e w_2^T \end{bmatrix}.$$

Thus, taking the second components of both sides of (7.32) gives,

$$-\alpha \pi_1^T H_{12} + \pi_2^T - \alpha \pi_2^T e w_2^T = (1 - \alpha)v^T.$$

Then solving for π_2^T and noting that the condition, $\pi_1^T e + \pi_2^T e = 1$ must be satisfied, where $\pi_1 e$ can be written as $\|\pi_1\|_1$, we arrive at the expression (7.27) for π_2. $\qquad\square$

7.3 Improving the PageRank

We now focus on results in [1,2], in which the authors let the damping factor, α be denoted by c and consider the Google matrix given by (7.3).

Theorem 7.10 *(Lemma 2 in [2]) The PageRank of a webpage increases if it receives an inlink from another page.*

Proof. We consider two webpages, i and j, and study the change in the PageRank of site j if page i adds a new link to it. Since $\pi_j = \frac{1}{\mu_j}$ (see the Appendix), we aim to determine μ_j and for this purpose define the probabilities,

$$f_{ij}^{(m)} := P\left(X_m = j, \text{ for } 1 \leq k < m, X_k \neq i, j | X_0 = i\right),$$

$$g_{ij}^{(m)} := P\left(X_m = j, \text{ for } 1 \leq k < m, X_k \neq j | X_0 = i\right).$$

In words, $f_{ij}^{(m)}$ is the probability of the chain arriving at page j after m steps given that it started at state i and in its path did not visit neither state i nor state j. If it is allowed for the Markov chain to visit state i along the way, then this probability is denoted by $g_{ij}^{(m)}$. Similarly, $f_{jj}^{(m)}$ denotes the probability of the chain being at state j after m steps given that it started at state j and did not go through either state i or state j.

There are two ways for the chain to start from state j and come back to it. One is without visiting state i or j in the path, which is presented by $f_{jj}^{(m)}$ and another is to go to state i and then from state i to state j, which is formulated by $f_{ji}^{(m)} + g_{ij}^{(m)}$. Since μ_j is the average number of steps from j to j, then

$$\mu_j = \sum_{m=1}^{\infty} m f_{jj}^{(m)} + \sum_{m=1}^{\infty} m f_{ji}^{(m)} + \sum_{m=1}^{\infty} m g_{ij}^{(m)}.$$

Now we suppose a link is added from site i to site j. Since $f_{jj}^{(m)}$ and $f_{ji}^{(m)}$ do not visit state i, then they remain the same. Adding a link from i to j would make the path described by $g_{ij}^{(m)}$ much shorter, causing the third term to decrease, which in turn increases the PageRank. $\qquad\square$

For the theorem below we need the Laurent series expansion of Markov chains defined as follows. For more details we refer the reader to Appendix A of [28]. For the transition matrix, P of a Markov chain, let,

$$P^* = \lim_{N \to \infty} \frac{1}{N} \sum_{n=0}^{N-1} P^n,$$

which exists if the state space is finite (see Theorem A.6 in [28]). In addition, if P is recurrent and irreducible, then $P^* = e\pi^T$, where π is the stationary distribution of P. Let,

$$H_p := (I - P + P^*)^{-1}(I - P^*),$$

which for aperiodic Markov chains becomes,

$$H_p = \lim_{N \to \infty} \left(\sum_{k=0}^{N-1} P^k - NP^* \right).$$

Then for a constant ρ such that $0 < \rho < \sigma(I-P)$, where $\sigma(I-P)$ is the spectral radius of $I-P$, we have,

$$(\rho I + (I - P))^{-1} = \rho^{-1} P^* + \sum_{n=0}^{\infty} (-\rho)^n H_p^{n+1}, \tag{7.33}$$

referred to as the Laurent series expansion of the Markov chain. We also use the following notation. Let m_{ij} denote the average time the Markov chain takes to go from state i to state j. It satisfies,

$$m_{ij} = \frac{h_{jj} - h_{ij}}{\pi_j}, \tag{7.34}$$

where h_{ij} is the ij^{th} entry of the hyperlink matrix, H. Also we let, $Z := (I - cP)^{-1}$ and note that,

$$\pi = \frac{1-c}{n} e^T Z, \tag{7.35}$$

which is obtained by substituting $\widetilde{P} = cP + \frac{1-c}{n} ev^T$ in $\pi \widetilde{P} = \pi$, where $\pi e = 1$, since here π is thought of as a probability row vector.

Theorem 7.11 (*Theorems 3.1 and 4.1 in [1]*) *Suppose the damping factor, c is relatively close to 1. If webpage 1 creates k_1 outgoing links with one link to each webpage in the set $M = \{2, ..., k_1 + 1\}$, then the new PageRank vector is given by,*

$$\widetilde{\pi}_1 = \frac{\pi_1}{1 - \frac{k_1}{k}\left(1 + \frac{c}{k_1}\sum_{i=2}^{k_1+1} z_{i1} - z_{11}\right)}, \tag{7.36}$$

$$\widetilde{\pi}_j = \pi_j + \pi_1 \frac{\frac{k_1}{k}\left(\frac{c}{k_1}\sum_{i=2}^{k_1+1} z_{ij} - z_{1j}\right)}{1 - \frac{k_1}{k}\left(1 + \frac{c}{k_1}\sum_{i=2}^{k_1+1} z_{i1} - z_{11}\right)}, \quad for\ j \in M, \tag{7.37}$$

which implies that the PageRank of page 1 increases if $m_{11} > 1 + \frac{1}{k_1}\sum_{i=2}^{k_1+1} m_{i1}$ and the PageRank of a page $j \in M$ with $j \neq 1$ increases if $m_{1j} > 1 + \frac{1}{k_1}\sum_{i=2, i\neq j}^{k_1+1} m_{ij}$.

Proof. With the added links, let $\widetilde{P}$ be the new hyperlink matrix with the change given by $e_1 u^T$. That is let $\widetilde{P} = P + e_1 u^T$, where,

$$u^T = \frac{1}{k} \sum_{i=2}^{k_1+1} e_i^T - \frac{k_1}{k} p_1^T, \tag{7.38}$$

where e_i is the column vector with all zero entries except the ith entry being 1 and p_i is the ith column of P. Similar to (7.20), by the Sherman-Morrison formula (7.18) we have,

$$(I - c\widetilde{P})^{-1} = Z + c\frac{Ze_1 u^T Z}{1 - cu^T Ze_1}. \tag{7.39}$$

Since by (7.35), $\pi_j = \frac{1-c}{n} e^T (I - cP_j)^{-1}$, then multiplying both sides of (7.39) from the left by $\frac{1-c}{n} e^T$ produces,

$$\widetilde{\pi} = \pi + \pi_1 \frac{cu^T Z}{1 - cu^T Ze_1}.$$

Observe that $e_j^T A$ and Ae_j give the jth row and j^{th} column of A, respectively and $e_i^T Ae_j$ is the entry a_{ij}. Then,

$$\widetilde{\pi}_1 = \pi_1 + \pi_1 \frac{cu^T Ze_1}{1 - cu^T Ze_1} = \pi_1 \left(\frac{1}{1 - cu^T Ze_1} \right),$$

$$\widetilde{\pi}_j = \pi_j + \pi_1 \frac{cu^T Ze_j}{1 - cu^T Ze_1}.$$

Using the expression for u^T given in (7.38), we obtain,

$$cu^T Ze_j = \frac{k_1}{k} \left(\frac{c}{k_1} \sum_{i=2}^{k_1+1} z_{ij} - cp_1^T Ze_j \right). \tag{7.40}$$

Notice that $cP = I - (I - cP)$ and multiplying it from the right by $(I - cP)^{-1}$ gives, $cPZ = Z - I$. Furthermore, since a_1 denotes the first column of matrix A, we obtain for the row vector p_1^T, $cp_1^T Z = z_1^T - e_1^T$, implying that the jth column of both sides are the same as follows,

$$cp_1^T Ze_j = z_1^T e_j - e_1^T e_j = z_{1j} - e_1^T e_j. \tag{7.41}$$

Substituting (7.41) in (7.40) we obtain (7.36) and (7.37) by noting that $e_1^T e_j = 1$ when $j = 1$ and 0 otherwise.

Based on (7.36) and (7.37), we next determine conditions needed to ensure an increase in the PageRank of webpages 1 and $j \in M$. Let $c = \frac{1}{\rho+1}$, with $\rho > 0$ being a constant sufficiently close to zero. Then applying the Laurent series expansion given by (7.33) and noting that here the chain is recurrent and irreducible we obtain,

$$Z = \left(I - \frac{1}{1+\rho}P\right)^{-1} = (1+\rho)\,(\rho I - (P - I))^{-1} \tag{7.42}$$

$$= (1+\rho)\left(\frac{1}{\rho}P^* + H + \sum_{k=1}^{\infty}\rho^k(-1)^k H^{k+1}\right) = (1+\rho)\left(\frac{1}{\rho}e\pi + H + O(\rho)\right),$$

where we used the fact that $\left(\frac{1}{k}A\right)^{-1} = kA^{-1}$. To prove an increase in the PageRank of webpage 1, we need $\hat{\pi}_1 > \pi_1$, which by using the expression (7.36) for $\hat{\pi}_1$ simplifies to the condition,

$$1 + \frac{c}{k_1}\sum_{i=2}^{k_1+1} z_{i1} - z_{11} > 0.$$

By (7.42),

$$z_{i1} = (1+\rho)\left(\frac{1}{\rho}\pi_1 + h_{i1} + O(\rho)\right),$$

and since (7.34) implies, $h_{i1} = h_{11} - m_{i1}\pi_1$, we find,

$$1 + \frac{c}{k_1}\sum_{i=2}^{k_1+1} z_{i1} - z_{11} = 1 + \frac{1}{k_1}\sum_{i=2}^{k_1+1}(h_{i1}) - h_{11} - \pi_1 + O(\rho)$$

$$= 1 + \frac{1}{k_1}\sum_{i=2}^{k_1+1}(h_{11} - m_{i1}\pi_1) - h_{11} - \pi_1 + O(\rho)$$

$$= 1 - \pi_1\left(1 + \frac{1}{k_1}\sum_{i=2}^{k_1+1} m_{i1}\right),$$

where we applied the fact that $\sum_{i=a}^{b} k = k(b-a+1)$ and set $\rho \to 0$. Therefore, we need,

$$1 - \pi_1 \left(1 + \frac{1}{k_1} \sum_{i=2}^{k_1+1} m_{i1} \right) > 0,$$

which is exactly the condition given in the theorem by noting that $m_{11} = \frac{1}{\pi_1}$.

Similarly, to show an increase in the PageRank of webpage j in M, we need, $\hat{\pi}_j > \pi_j$, which by the expression in (7.37) becomes the condition,

$$\frac{c}{k_1} \sum_{i=2}^{k_1+1} z_{ij} - z_{1j} > 0,$$

since, $\frac{k_1}{k} > 0$. Following the same notation as above,

$$\frac{c}{k_1} \sum_{i=2}^{k_1+1} z_{ij} - z_{1j} = \frac{1}{k_1} \sum_{i=2}^{k_1+1} h_{ij} - h_{1j} - \pi_j + O(\rho).$$

Breaking up the series and using (7.34) leads to,

$$\frac{1}{k_1} \sum_{i=2, i \neq j}^{k_1+1} \left(h_{jj} - m_{ij} \pi_j - h_{jj} + m_{1j} \pi_j - \pi_j \right) + \frac{1}{k_1} \left(m_{1j} \pi_j - \pi_j \right) + O(\rho)$$

$$= \frac{\pi_j}{k_1} \left(\sum_{i=2, i \neq j}^{k_1+1} \left(m_{1j} - m_{ij} - 1 \right) + \left(m_{1j} - 1 \right) \right),$$

Note that since the series above excludes $i \neq j$ and $j \in \{2, ..., k_1 + 1\}$, then,

$$\sum_{i=2, i \neq j}^{k_1+1} a = a \left((k_1 + 1) - 2 + 1 - 1 \right) = a(k_1 - 1).$$

Hence we arrive at,

$$= \frac{\pi_j}{k_1} \left(m_{1j}(k_1 - 1) - \sum_{i=2, i \neq j}^{k_1+1} m_{ij} - (k_1 - 1) + m_{1j} - 1 \right)$$

$$= \frac{\pi_j}{k_1} \left(k_1 m_{1j} + \sum_{i=2, i \neq j}^{k_1+1} m_{ij} - k_1 \right),$$

which is positive if the condition given in the theorem holds.   □

If we group webpages of the same content type together, then an inappropriate link is a link from a state in group i to a state in group j, where not many pages in group j are linked back to group i. These groups are called web communities. If we assume that the worldwide web consists of only two web communities, then the hyperlink matrix without any inappropriate link is,

$$P = \begin{array}{c} \\ G1 \\ G2 \end{array} \begin{array}{cc} G1 & G2 \\ \begin{bmatrix} P_1 & 0 \\ 0 & P_2 \end{bmatrix}, \end{array}$$

and it changes to,

$$\hat{P} = \begin{array}{c} \\ G1 \\ G2 \end{array} \begin{array}{cc} G1 & G2 \\ \begin{bmatrix} P_1 & 0 \\ R & \hat{P}_2 \end{bmatrix}, \end{array} \qquad (7.43)$$

if there are inappropriate links from community 2 to community 1. Recall that PageRank is defined as the unique stationary distribution, π such that,

$$\pi P = \pi, \text{ and } \pi e = 1, \qquad (7.44)$$

where in [1,2] they let π be a row probability vector, denote e as $\underline{1}$ and $E = ee^T$. To be consistent, we keep the notation e for a column vector of all ones. Replacing P in (7.35) with $\hat{P}$ given in (7.43) representing the web with inappropriate links, we arrive at,

$$\hat{\pi}(I - c\hat{P}) = \frac{1-c}{n}e^T.$$

Since it is assumed that there are only two web communities, then,

$$\begin{bmatrix} \hat{\pi}_1 & \hat{\pi}_2 \end{bmatrix} \begin{bmatrix} I - \begin{bmatrix} cP_1 & 0 \\ cR & c\hat{P}_2 \end{bmatrix} \end{bmatrix} = \frac{1-c}{n}e^T,$$

which leads to,

$$\begin{bmatrix} \hat{\pi}_1 & \hat{\pi}_2 \end{bmatrix} \begin{bmatrix} I - cP_1 & 0 \\ -cR & I - c\hat{P}_2 \end{bmatrix} = \begin{bmatrix} \hat{\pi}_1(I - cP_1) - cR\hat{\pi}_2 & \hat{\pi}_2(I - c\hat{P}_2) \end{bmatrix} = \frac{1-c}{n}e^T,$$

and solving for $\hat{\pi}_1$ and $\hat{\pi}_2$ gives the new PageRanks for web communities 1 and 2 as follows,

$$\hat{\pi}_1 = cR\hat{\pi}_2(I - cP_1)^{-1} + \frac{1-c}{n}e^T(I - cP_1)^{-1} = cR\hat{\pi}_2(I - cP_1)^{-1} + \pi_1,$$

$$\hat{\pi}_2 = \frac{1-c}{n}e^T(I - c\hat{P}_2)^{-1}.$$

Observe that since $I - cP_1$ is an M-matrix, then $(I - cP_1)^{-1}$ is positive, implying that the PageRank of the first web community increased by $cR\hat{\pi}_2(I - cP_1)^{-1}$, which complies with the fact that the web community 1 received new inlinks. Some web developers tend to increase their website's rank by creating a group of websites called spam farms to link to their target page. For this purpose, they create advertisements and purchase expired domains and fill them with links pointing to their target page. In [14], Z. Gyöngyi and H. Garcia-Molina explain various methods spammers use to gain undeserved enhancements of their PageRank and discuss ways that they apply to conceal their tactics. Some spammers cooperate with one another by exchanging links between each other's spam farms.

Algorithms have been developed in the literature aimed at detecting link spamming. For example, in [4], A. Benczúr, K. Csalogány, T. Sarlós, and M. Uher present an algorithm that measures a webpage's level of spam referred to as SpamRank. Another algorithm designed to detect link spamming is offered in [3] by L. Becchetti, et. al.

One may also reason mathematically, when link spamming has played a role in a webpage's PageRank. In the context of web communities, if the number of links from an unrelated and lower ranked web community i going to a page in web community j, is less than the number of links connecting pages in community j, then link spam has likely occurred. In addition, the second eigenvalue of the Google matrix can be used to detect link spamming. It is known that the convergence rate of the power method depends on the two largest eigenvalues, λ_1 and λ_2 of the Google matrix. More precisely, it is equivalent to the rate of convergence of $\left|\frac{\lambda_2}{\lambda_1}\right|^k \to 0$ and since $\lambda_1 = 1$, we have $|\lambda_2|^k \to 0$. By Theorem 7.4 in Section 7.1, this rate translates to $c^k \to 0$, that is $\alpha^k \to 0$ based on our original notation for the damping factor. Since Google assigns $\alpha = 0.85$ to each webpage, then if the rate of convergence of $|\lambda_2|^k \to 0$ is considerably different than that of $(0.85)^k \to 0$, then one may suspect link spamming.

7.4 Appendix

7.4.1 Stochastic Processes

Recall that a random variable, $X : \Omega \to \mathbb{R}$, is a function that randomly assigns a value to each element, ω in the set of possible events, Ω. We denote $(\Omega, \mathcal{F}, P)$

to be the probability space associated with the random variable, where P and $\mathcal{F}$ stand for its probability measure and σ-algebra of the events, respectively. σ-algebra is a family of subsets of events in Ω that contain Ω and is closed under complement, countable union and countable intersection. The precise definition of a random variable is a function, X, such that for all sets, B in the Borel σ-algebra of $\mathbb{R}$ (the σ-algebra generated by all open intervals in $\mathbb{R}$), $X^{-1}(B)$ is a subset of the σ-algebra $\mathcal{F}$, meaning that X is measurable. Furthermore, a filtration, $\{\mathcal{F}_t\}_{t \geq 0}$ on $(\Omega, \mathcal{F})$ is a family of sub-σ-algebras of $\mathcal{F}$ with the property that $\mathcal{F}_s \subset \mathcal{F}_t$ if $t \geq s$. A collection of random variables, $\{X_t\}_{t \geq 0}$ is called a stochastic process, where, $X_t = X(t, \omega)$ depends on index time, $t \in [0, \infty)$ and $\omega \in \Omega$. In the case that t is discrete, meaning it only takes values in natural numbers or zero, t is sometimes denoted by n and the process is called a discrete stochastic process.

7.4.2 Markov Process and Markov Chain

A stochastic process, $\{X_t\}_{t \geq 0}$ is called a Markov process if it satisfies the Markov property given below,

$$P\left(X_{t+s} \in A | \mathcal{F}_t\right) = P\left(X_{t+s} \in A | X_t\right), \tag{7.45}$$

where $t \geq s$, each $X_t : E \times [0, \infty) \to \mathbb{R}$ and A is a Borel set of the space E. The above property means that a future event only depends on the information on the present and not on the past. It can be read as the probability of X_{t+s} being in A given information up to time t, and the right hand side of (7.45) indicates that this probability only depends on information at time t. When the process is discrete, then the process is called a Markov chain and the Markov property becomes,

$$P\left(X_{n+1} = i | X_{t_0} = i_0, X_{t_1} = i_1, ..., X_n = i_n\right) = P\left(X_{n+1} = i | X_n = i_n\right), \tag{7.46}$$

where $n \in \mathbb{N}$. Hence, only the information from the present observation, X_n, matters.

7.4.3 Properties of Markov Chains

For every Markov chain, there is an associated transition matrix, P that is a row stochastic matrix and nonnegative, meaning that every entry is nonnegative. Each entry, p_{ij} in P gives the probability of the next step of the process being at state j, given that the current step is at state i. That is for every i, j in the set of states,

$$P_{ij} = P\left(X_1 = j | X_0 = i\right). \tag{7.47}$$

When each of the probabilities in a transition matrix does not depend on n, then the Markov chain is said to be homogenous. This translates to $P(X_{u+n} = j|X_0 = i) = P(X_u = j|X_0 = i)$ for a positive integer u.

We consider, for an example, the three-state graph below, in which the numbers next to edges represent the probability of going from one state to another noting the directions on the edges.

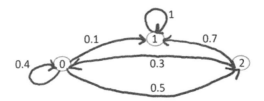

FIGURE 7.1
Example of Markov Chain.

For the above graph we have,

$$
P = \begin{array}{c} \\ 0 \\ 1 \\ 2 \end{array} \begin{array}{ccc} 0 & 1 & 2 \\ \left[\begin{array}{ccc} P_{00} & P_{01} & P_{02} \\ P_{10} & P_{11} & P_{12} \\ P_{20} & P_{21} & P_{22} \end{array} \right] \end{array} = \begin{array}{c} \\ 0 \\ 1 \\ 2 \end{array} \begin{array}{ccc} 0 & 1 & 2 \\ \left[\begin{array}{ccc} 0.4 & 0.1 & 0.5 \\ 0 & 1 & 0 \\ 0.3 & 0.7 & 0 \end{array} \right] \end{array}
$$

One can observe that the sum of the entries in each row is 1, which is the requirement of a transition matrix of a Markov chain. For two states, i and j, if there is an arrow from i to j and one from j to i, then we say that the states i and j communicate. That is i and j communicate if $i \leftrightarrow j$ meaning $P(X_1 = j|X_0 = i) > 0$ and $P(X_1 = i|X_0 = j) > 0$. For instance, in the above figure, states 0 and 2 communicate. If $P(X_1 = j|X_0 = j) = 1$, then state j is called an absorbing state. In figure above, 1 is an absorbing state.

7.4.4 Types of Markov Chains

If all the states in a Markov chain communicate, then the Markov chain is said to be irreducible. If there are more than one communication class, then the Markov chain is reducible. The Markov chain in Figure 7.1 is reducible since it is made up of three communication classes: $0 \leftrightarrow 2$, $1 \leftrightarrow 1$, $0 \leftrightarrow 0$.

A set of states, S in a Markov chain is called a closed subset if $P_{ji} = 0$ for every $i \in S$ and $j \notin S$. That is there is no transition from a state in S to a state outside of S. This property implies $\sum_{j \in S} P_{ij} = 1$ for every $i \in S$.

7.4.5 Computing the Transition Probabilities

The transition matrix gives the probability of going to state j at time $n = 1$, given that the chain is at the state i at time $n = 0$. To determine the state at later times, we use the following formula,

$$P\left(X_{n+k} = j | X_n = i\right) = P_{ij}^k, \tag{7.48}$$

for any positive integer, k. For example, setting $n = 0$ and $k = 1$, we obtain,

$$P\left(X_1 = j | X_0 = i\right) = P_{ij},$$

which agrees with equation (7.47). For $n = 0$ and $k = 2$ we have,

$$P\left(X_2 = j | X_0 = i\right) = P_{ij}^2,$$

and hence for the Markov chain in Figure 7.1,

$$
P^2 = \begin{array}{c} \\ 0 \\ 1 \\ 2 \end{array}
\begin{array}{ccc} 0 & 1 & 2 \end{array}
\left[
\begin{array}{ccc}
0.31 & 0.49 & 0.2 \\
0 & 1 & 0 \\
0.12 & 0.73 & 0.15
\end{array}
\right],
$$

from which we have for example, $P(X_2 = 1 | X_0 = 2) = 0.73$, $P(X_2 = 0 |$ $X_0 = 0) = 0.31$ and $P(X_2 = 0 | X_0 = 2) = 0.12$. Equation (7.48) implies,

$$P\left(X_m = j | X_0 = i\right) = P_{ij}^m,$$

giving the probability of going to state j after time m (after m steps), if initially at $n = 0$ the process starts at state i. Note that based on (7.48), we also obtain,

$$P\left(X_{n+1} = j | X_n = i\right) = P_{ij},$$

by setting $k = 1$, which confirms the fact that the Markov chain is time-homogenous. Hence,

$$P_{ij} = P\left(X_1 = j | X_0 = i\right) = P\left(X_{n+1} = j | X_n = i\right).$$

If the Markov chain starts at state i and after finitely many steps returns to i, then state i is called recurrent. This translates to,

$$P_{ii} = P\left(X_n = i \text{ for some } n \geq 1 | X_0 = i\right) = 1.$$

If this probability is less than 1, then the state is transient. Since a Markov chain acts as a cycle moving through the nodes, then if i is a recurrent state, then $\sum_{n=1}^{\infty} (P^n)_{ii} = \infty$ and if it is transient, $\sum_{n=1}^{\infty} (P^n)_{ii} < \infty$. It is intuitively clear that any state that communicates with a recurrent state, is itself recurrent. Also if a Markov chain has finitely many states, then it must have at least one recurrent state.

One may follow the arrows on the graph and determine in how many steps the process can move out of state i and return to it. There can be more than one path available. The greatest common divisor of the number of steps taken by different possible paths is referred to as the period of state i and when this number is 1, then state i is said to be aperiodic. Notice that an absorbing state, which has a loop of probability 1, is recurrent and aperiodic, since the loop offers a path with $n = 1$, making the greatest common divisor of the set be 1. In Figure 7.1, for state 0, one of the paths is the loop, for which $n = 1$ and the other path is $0 \rightarrow 2 \rightarrow 0$ giving $n = 2$. Thus,

$$\left\{ n \geq 1 : (P^n)_{00} > 0 \right\} = \{1, 2\}.$$

Similarly,

$$\left\{ n \geq 1 : (P^n)_{11} > 0 \right\} = \{1\} \quad \text{and} \quad \left\{ n \geq 1 : (P^n)_{22} > 0 \right\} = \{2\},$$

implying that 0 and 1 are aperiodic and state 2 has period 2.

We say that a matrix A is nonnegative if all its entries are nonnegative and it is primitive (also called regular) if there exists a positive integer, k such that $A^k > 0$. It is well-known that a nonnegative matrix is primitive if and only if it is irreducible and aperiodic, where irreducibility refers to the matrix A not being reducible by means of row operations.

7.4.6 Stationary Probability Distribution

Starting at state i, it is sometimes useful to determine the probability of arriving at state j after a large number of steps. That is, studying the long term behavior of the process, which is determined by the limiting distribution,

$$\pi_j = \lim_{n \to \infty} P\left(X_n = j | X_0 = i\right),$$

if it exists. The vector, $\pi^T = [\pi_1 ... \pi_n]$ is called a stationary probability distribution if $\sum_{i=1}^{n} \pi_i = 1$ and $\pi = \pi P$.

An irreducible, homogeneous Markov chain with finite state space has a unique stationary distribution given by,

$$\pi_j = \frac{1}{\mu_j},$$ (7.49)

where, $\mu_j := \mathbb{E}\left(T_j^r : X_0 = j\right)$ is the average time it takes for the chain starting from state j to return to j. In the definition of μ_j, the stopping time T_j^r is the rth time that the chain visits state j (see Theorems 3.2.6–3.2.8 in [7]).

References

1. K. Avrachenkov and N. Litvak (2006). The effect of new links on Google PageRank. *Stoch. Models.* vol. 22, 319–331.

2. K. Avrachenkov and N. Litvak (2004). *Decomposition of the Google PageRank and Optimal Linking Strategy.* INRIA Sophia Antipolis. Doctoral Dissertation.

3. L. Becchetti, C. Castillo, D. Donato, R. Baeza-Yates, and S. Leonardi (2008). Link analysis for web spam detection. *ACM Transaction.* vo. 21, no. 2, 1–42.

4. A. Benczúr, K. Csalogány, T. Sarlós, and M. Uher (2005). SpamRank-fully automatic link spam detection. *Proceedings of the 1ˢᵗ International Workshop on Adversarial Information Retrieval on the Web.*

5. P. Berkhin (2005). A survey on PageRank computing. *Internet Math.* vol. 2, no. 1, 73–120.

6. M. Bianchini, M. Gori and F. Scarselli (2005). Inside PageRank. *ACM Trans. Internet Tech.* vol. 5, no. 1, 92–128.

7. P. Brémaud (2010).*Markov Chains: Gibbs Fields, Monte Carlo Simulation and Queues: Second Edition.* Springer Texts in Applied Mathematics, vol. 31, Cham, Switzerland.

8. C. Brezinski and M. Redivo-Zaglia (2006). The PageRank vector: properties, computation, approximation and accelaration. *SIAM J. Matrix Anal. Appl.* vol. 28, no. 2, 551–575.

9. S. Brin and L. Page (1998). The anatomy of a large-scale hypertextual web search engine. *Compt. Net. ISDN Syst.* vol. 30, no. 1-7, 107–117.

10. S. Brin and L. Page (1999). The PageRank citation ranking: bringing order to the web. *Standford InfoLab.* Technical Report.

11. K. Bryan and T. Leise (2006). The $\$25,000,000,000$ eigenvector: the linear Algebra behind Google. *SIAM Review.* vol. 48, no. 3, 569–581.

12. G. Corso, A. Gulli, and F. Romani (2005). Fast PageRank computation via a sparse linear system. *Internet Math.* vol. 2, no. 3, 251–273.

13. J. Ding and A. Zhou (2007). Eigenvalues of rank-one updated matrices with some applications. *Applied Mathematics Letters*. vol. 20, 1223–1226.

14. Z. Gyöngyi and H. Garcia-Molina (2005). Web spam taxonomy. *Proceedings of the 1st International Workshop on Adversarial Information Retrieval on the Web*. AIRWeb.

15. T. Haveliwala and S. Kamvar (2003). The second eigenvalue of the Google matrix. *Technical Report*. Standford InfoLab.

16. M. Iosifescu (1980). *Finite Markov Processes and their Applications*. John Wiley & Sons. New York.

17. I. Ipsen and T. Selee (2007). PageRank computation with special attention to dangling nodes. *SIAM J. Matrix Anal. Appl.* vol. 29, no. 4, 1281–1296.

18. D. Isaacson and W. Madsen (1976). *Markov Chains: Theory and Applications*. John Wiley & Sons. New York.

19. S. Kamvar, T. Haveliwala, and G. Golub (2004). Adaptive methods for the computation of PageRank. *Linear Alg. Appl.* vol. 386, 51–65.

20. A. Langville and C. Meyer (2004). Deeper inside PageRank. *Internet Mathematics*. vol. 1, no. 3, 335–380.

21. A. Langville and C. Meyer (2005). A survey of eigenvector methods for web information retrieval. *SIAM Review*. vol. 47, no. 1, 135–161.

22. A. Langville and C. Meyer (2006). A reordering for the PageRank problem. *Soc. Indus. Appl. Math.* vol. 27, no. 6, 2112–2120.

23. A. Langville and C. Meyer (2011). *Google's PageRank and Beyond: The Science of Search Engine Rankings*. Princeton University Press, New Jersey.

24. C. Lee, G. Golub, S. Zenios (2008). A two-stage algorithm for computing PageRank and multistage generalizations. *Internet Math.* vol. 4, no. 4., 299–327.

25. B. Noble and J. Daniel (1988). *Applied Linear Algebra: Third Edition*. Prentice Hall, New Jersey.

26. R. J. Plemmons (1977). M-Matrix characterizations. I-nonsingular M-Matrices. *Linear Alg. Appl.* vol. 18, no. 2, 175–188.

27. N. Privault (2013). *Understanding Markov Chains: Examples and Applications*. Springer Undergraduate Mathematics Series, London, United Kingdom.

28. M. Puterman (2005). *Markov Decision Processes: Discrete Stochastic Dynamic Programming*. John Wiley & Sons Inc., New Jersey.

8

Stochastic Navier-Stokes Equations in Fluids

The study of the heat and wave equations lie in the core of graduate level courses in partial differential equations (PDEs). Originating from the wave equation, fluid models are extensively studied in the research community. Here we mainly concentrate on the Navier-Stokes equation, which is given by,

$$\frac{\partial u}{\partial t} - \nu \Delta u + (u \cdot \nabla)u + \nabla p = f, \qquad (8.1)$$

where $u(t,x)$ represents the velocity of the fluid at time t and position x, $p(t,x)$ stands for its pressure at (t,x) and $f(t,x)$ is the external force. The constant $\nu > 0$ denotes the viscosity, which measures the thickness of the fluid. For example, oil has a higher viscosity than water. If the divergence of u is assumed to be zero, that is if $\text{div}(u) = \nabla \cdot u = 0$, then the above equation is referred to as the incompressible Navier-Stokes equation. If this assumption is not made, then the flow is considered to be compressible flow such as gas instead of liquid. Most of the work in the literature regarding the Navier-Stokes equations has been on the two-dimensional incompressible equations; however, the study on the three-dimensional Navier-Stokes equation and the equation in case of compressible flow are beginning to take shape. The compressible Navier-Stokes equation is precisely characterized by the following system,

$$\partial_t \rho + \text{div}(\rho u) = 0, \qquad (8.2)$$

$$\partial_t(\rho u) + \text{div}(\rho u \otimes u) + \nabla p(\rho) = \mu \Delta u + (\lambda + \mu)\nabla \text{div}(u) + \rho f, \qquad (8.3)$$

where equations (8.2) and (8.3) represent the balance of mass and momentum, respectively. The product, $u \otimes v$, between two n-dimensional vectors, u and v, is called a tensor product and forms the matrix, $A := u \otimes v$ with entries determined by $a_{ij} = u_i v_j$ where $i,j = 1,2,...,n$. Sometimes equation (8.3) is written as,

$$\partial_t(\rho u) + \text{div}(\rho u \otimes u) + \nabla p(\rho) = \text{div}(\mathbb{S}(u)) + \rho f,$$

where, $\mathbb{S}(u)$ is called the stress tensor satisfying $\text{div}(\mathbb{S}(u)) = \mu \Delta u + (\lambda + \mu)\nabla \text{div}(u)$. In incompressible Navier-Stokes equation, the pressure only

DOI: 10.1201/9781003299073-8

depends on the position and time; whereas, in compressible system it is also a function of the density. In addition, in compressible case there are two viscosity coefficients, μ and η, called the shear and bulk viscosity coefficient, respectively. In the above system, $\lambda := \eta - \frac{2}{3}\mu$ was used.

In this chapter we focus on the stochastic Navier-Stokes equations. Stochastic equations are those that incorporate the "noise" in the system by adding an extra term. "Noise" means anything that has an effect on the behavior of the system that has not been completely modeled in the equation. For example, the "noise" in laser light propagation through air might be air resistance, temperature or very small particles such as dust that can affect the direction of the laser and its intensity as it propagates. The noise term is usually given by $\sigma(x, u(x))dW$, where W is a Wiener process, also referred to as the Brownian motion. In the Appendix, we have offered a short introduction to Brownian motion and other commonly studied processes in the literature.

The above Navier-Stokes equations is referred to as deterministic equation. The stochastic counterpart, for example for the incompressible equation (8.1), is given by,

$$\frac{\partial u}{\partial t} - \nu\Delta u + (u \cdot \nabla)u + \nabla p = f + \sigma(t, u(t))\frac{dW}{dt}, \qquad (8.4)$$

where W is a Wiener process. We refer to $\sigma(t, u(t))\frac{dW}{dt}$ as the noise term. When σ depends on the process, $\{u(t)\}_{t\geq 0}$, as in (8.4), we call the noise multiplicative noise and when σ only depends on t and x and not on the process then the noise is called additive noise. Hence, typically results are first proved for the equation with additive noise and then extended to the case with multiplicative noise. In the study of the asymptotic behavior of stochastic partial differential equations (SPDEs), the noise term is often multiplied by a function of $\varepsilon > 0$ and setting $\varepsilon \to 0$ gives the form of the solution as the noise in the system is set to diminish to zero.

The integral $\int f(t)dW(t)$ is called a stochastic or an Itô integral. Brownian motion is continuous but nowhere differentiable and thus, the usual rules of integration can not be applied to stochastic integrals. The main tool in stochastic analysis is the Itô formula and is applied as follows. In the one-dimensional noise setting, let $\{X_t\}_{t\in[0,T]}$ be the process, called an Itô process, formed by solutions to,

$$dX_t = \mu(t)dt + \sigma(t)dB(t), \qquad (8.5)$$

where B is a one-dimensional Brownian motion. Then for a twice continuously differentiable function, $g(t, x)$ (denoted by $g \in C^2([0, \infty) \times \mathbb{R})$), the Itô formula gives,

$$dg(t, X(t)) = \frac{\partial g}{\partial t}dt + \frac{\partial g}{\partial x}dX_t + \frac{1}{2}\frac{\partial^2 g}{\partial x^2}(dX_t)^2, \tag{8.6}$$

with the convention, $dt \cdot dt = dt \cdot dB_t = 0$ and $dB_t \cdot dB_t = dt$.

For example, suppose we have the process $\{X_t\}_t$ given by (8.5) and we wish to find X_t^2. Then $g(x) = x^2$ and since g does not depend on t, then the first term on the right of (8.6) will be dropped and using $g'(x) = 2x$ and $g''(x) = 2$, we obtain,

$$dX_t^2 = 2X_t dX_t + \frac{1}{2}2(dX_t)^2.$$

To determine $(dX_t)^2$, note that by the convention above,

$$(dX_t)^2 = (\mu(t)dt + \sigma(t)dB(t))^2$$

$$= \mu(t)^2(dt)^2 + 2\mu(t)\sigma(t)dt dB(t) + \sigma(t)^2(dB(t))^2 = \sigma(t)^2 dt.$$

Hence,

$$dX_t^2 = 2X_t\mu(t)dt + 2X_t\sigma(t)dB_t + \sigma(t)^2 dt,$$

yielding,

$$X_t^2 = 2\int_0^t X_s\mu(s)ds + 2\int_0^t X_s\sigma(s)dB_s + \int_0^t \sigma(s)^2 ds.$$

In the infinite dimensional setting, such as in a Hilbert space, H, the following form of Itô formula is used (see for example Section 4.4 in [11]). Let $F : [0, T] \times H \to \mathbb{R}$ be in $\mathcal{C}^2([0, T] \times H)$, then for Itô process given by,

$$X(t) = X(0) + \int_0^t \varphi(s)ds + \int_0^t \Phi(s)dW(s),$$

the Itô formula is defined using the inner product in H as,

$$F(t, X(t)) = F(0, X(0)) + \int_0^t \langle F_t(s, X(s)), \varphi(s)\rangle \, ds + \int_0^t \langle F_x(s, X(s)), \varphi(s)\rangle \, ds$$

$$+ \int_0^t \langle F_x(s, X(s)), \Phi(s)dW(s)\rangle$$

$$+ \frac{1}{2}\int_0^t \mathrm{Tr}\left(F_{xx}(s, X(s))\left(\Phi(s)Q^{1/2}\right)\left(\Phi(s)Q^{1/2}\right)^*\right)ds. \tag{8.7}$$

In the last term, noting that W is a Gaussian process, Q is its covariance operator. Furthermore, Q^* denotes the adjoint operator and for a complete orthonormal system (CONS), $\{e_n\}_n$ in H, the trace of Q is defined as,

$$\text{Tr}(Q) := \sum_{n=1}^{\infty} \langle Qe_n, e_n \rangle .$$

For instance, in the case of the Navier-Stokes Equation (8.4), with $H_0 := Q^{1/2}H$ we have for $u, v \in H$,

$$< u, v >_{H_0} := \left(Q^{-1/2}u, Q^{-1/2}v \right).$$

Letting $L_Q(H_0; H)$ be the space of operators, $S : H_0 \to H$ that satisfy,

$$\|S\|_{L_2} := \left(\sum_{j=1}^{\infty} \|Se_j\|_H^2 \right)^{1/2} < \infty, \tag{8.8}$$

for a CONS, $\{e_j\}_j$ in H, the norm in $L_Q(H_0; H)$ is defined as,

$$\|S\|_{L_Q} := \sqrt{\text{Tr}(SQS^*)}. \tag{8.9}$$

We note that the norm defined in (8.8) is referred to as the Hilbert Schmidt norm and operators, φ with $\|\varphi\|_{L_2} < \infty$ are said to be Hilbert Schmidt operators. Now applying the Itô formula (8.7) to (8.4) with $F(X(t)) = |X(t)|^2$, yields,

$$|u(t)|^2 = |u(0)|^2 + 2\nu \int_0^t \langle \Delta u(s), u(s) \rangle \, ds - 2 \int_0^t \langle (u(s) \cdot \nabla)u(s), u(s) \rangle \, ds$$

$$- 2 \int_0^t \langle \nabla p(s), u(s) \rangle \, ds + 2 \int_0^t \langle f(s), u(s) \rangle \, ds$$

$$+ 2 \int_0^t \langle \sigma(s, u(s))dW(s), u(s) \rangle + \int_0^t \|\sigma(s, u(s))\|_{L_Q}^2 \, ds. \tag{8.10}$$

We further discuss in Section 8.1 how to simplify the terms above and we state the conditions that are typically applied.

We note that in some applications, the Stratonovich integral, defined as follows, is used instead of the Itô integral. For two Itô processes,

$$dX_t = b_1(t)dt + \sigma_1(t)dW(t),$$
$$dY_t = b_2(t)dt + \sigma_2(t)dW(t),$$

the Stratonovich integral is denoted by $\int X_s \circ dY_s$ and may be written in terms of an Itô integral as,

$$\int X_s \circ dY_s = \int X_s dY_s + \frac{1}{2} \int (dX_s)(dY_s),$$

simplifying to,

$$\int X_s \circ dY_s = \int X_s b_2(s) ds + \int X_s \sigma_2(s) dW(s) + \frac{1}{2} \int \sigma_1(s) \sigma_2(s) ds.$$

As for the types of noise that are typically used for SPDEs in the infinite dimensional setting, Q-Wiener process and cylindrical Brownian motion are the most common ones. As above, denoting Q to be the covariance operator of the Gaussian process, W, we have that W is a Q-Wiener process taking values in a Hilbert space, H, if for a CONS, $\{e_j\}_{j \geq 1}$ satisfying, $Q e_j = \lambda_j e_j$ and an infinite sequence of independent, standard, one-dimensional Brownian motions, $\{\beta_j\}_{j \geq 1}$,

$$W(t) := \sum_{j=1}^{\infty} \sqrt{\lambda_j} \beta_j(t) e_j. \tag{8.11}$$

Note that λ_j is the jth eigenvalue of Q corresponding to the eigenvector e_j. On the other hand, if for an CONS, $\{e_j\}_j$ in H and an infinite sequence of standard, independent and here real-valued Brownian motions, $\{\beta_j\}_{j \geq 1}$,

$$W_t(h) := \sum_{j=1}^{\infty} \beta_j(t) \langle e_j, h \rangle, \tag{8.12}$$

then W is called a cylindrical Brownian motion. For instance, if W in (8.4) is a cylindrical Brownian motion, then the term $2 \int_0^t \langle \sigma(s, u(s)) dW(s), u(s) \rangle$ in the Itô formula (8.10) becomes,

$$2 \sum_{j=1}^{\infty} \int_0^t \left(\sigma(s, u(s)) e_j, u(s) \right) d\beta_j(s).$$

We now give a description on different types of solutions that are used in SPDEs. For a stochastic equation,

$$u(t) = u(0) + \int_0^t b(s) ds + \int_0^t \sigma(s) dW(s), \tag{8.13}$$

a weak in PDE solution aligns with the usual definition that for every continuous, bounded function, $\varphi(t)$, the following equation holds,

$$(u(t), \varphi(t)) = (u(0), \varphi(0)) + \int_0^t (b(s), \varphi(s)) \, ds + \int_0^t (\sigma(s)dW(s), \varphi(s)). \quad (8.14)$$

If one keeps the same Wiener process, W and filtration, $\{\mathcal{F}_t\}_t$ as in the original
equation (8.13), then the solution given by (8.14) is weak in PDE but it is said
to be strong in probability. On the other hand, if the filtration or the Wiener
process has to be changed for (8.14) to hold, then the solution is weak in
both probability and PDE sense. In infinite dimensional setting, solutions
that are weak in probability and PDE are also referred to as martingale
solutions.

Recall from Chapter 6, the mild solution to the nonlinear Schrödinger
equation is,

$$\psi(t) = U(t)\varphi + iv \int_0^t U(t - s)|\psi(s)|^{2\sigma} \psi(s)ds, \quad (8.15)$$

with semigroup, $U(t) = e^{it\Delta}$. To obtain the form of a mild solution, one
multiplies each term by the appropriate semigroup, $S(t)$, which is a one
parameter operator satisfying the properties, $S(0) = I$ (the identity map),
and $S(t+s) = S(t)S(s)$. Similar to (8.15), the mild solution to the more general
equation,

$$X(t) = X(0) + \int_0^t b(s)ds + \int_0^t \sigma(s, X(s))dW(s),$$

is

$$X(t) = S(t)X(0) + \int_0^t S(t - s)b(s)ds + \int_0^t S(t - s)\sigma(s, X(s))dW(s),$$

for a suitable semigroup formed based on the operators in the equation. For
example, for Navier-Stokes equation (8.4), the semigroup, $S(t) = e^{-t\Delta}$ is
generated by the Laplace operator in the equation and is applied to obtain
the mild solution,

$$u(t) = S(t)u(0) - \int_0^t S(t - s)(u(s) \cdot \nabla)u(s)ds - \int_0^t S(t - s)\nabla p(s)ds$$

$$+ \int_0^t S(t - s)f(s)ds + \int_0^t S(t - s)\sigma(s, u(s))dW(s).$$

Notice that $S(t)v := e^{-t\Delta v}$, thus the integral, $\int_0^t S(t - s)\Delta u(s)ds$ becomes zero. The mild solution for the deterministic equation (8.1) may be formed as above with the same semigroup. In general, using the semigroup as the test function, it is intuitively clear that if (X, W) is a weak solution, then it is a mild solution. For a comprehensive study of mild solutions and properties of semigroups, we recommend [27].

Recall that a local solution is one that satisfies the equation only on a specific interval of time. More precisely, $u(t)$ is a local strong solution to (8.13) if there exists a stopping time, τ such that P-a.s. for all $t \in [0, T]$,

$$u(t \wedge \tau) = u(0) + \int_0^{t\wedge\tau} b(s)ds + \int_0^{t\wedge\tau} \sigma(s)dW(s),$$

where, $a \wedge b := \min\{a, b\}$. Furthermore, $u(t)$ is a maximal solution, also referred to as maximal local solution, if there exists an increasing sequence of stopping times, $\{\tau_n\}_n$ such that for each n,

$$u(t \wedge \tau_n) = u(0) + \int_0^{t\wedge\tau_n} b(s)ds + \int_0^{t\wedge\tau_n} \sigma(s)dW(s),$$

and there exists a strictly positive stopping time, τ_∞ such that $\lim_{n\to\infty} \tau_n = \tau_\infty$ a.s. and

$$\limsup_{t\to\tau_\infty} |u(t)|_X = \infty \qquad \text{P-a.s. on } \{\tau_\infty \leq T\}.$$

To show that a maximal solution, $u(t)$ is indeed a global solution one proves that $\tau_\infty = T$. For examples of results on maximal solutions see for example, [3, 10, 13].

As for the uniqueness of solutions to stochastic equations, it is sometimes sufficient for the results to prove the pathwise uniqueness. The solution to a stochastic equation is pathwise unique if for two weak solutions, (X, W) and (X', W) on the same space, $(\Omega, \mathcal{F}, P)$ and filtration, $\{\mathcal{F}_t\}_{t\geq 0}$,

$$P\left(X_t = X'_t \text{ for every } t \geq 0\right) = 1.$$

That is, for each fixed t, $X_t = X'_t$. Observe that the solutions (X, W) and (X', W) as described are strong in probability and weak in PDE. Another notion is uniqueness in law, which is defined for two weak solutions that have different filtration and Wiener process. Namely, the uniqueness in law holds for a stochastic equation if for any two weak solutions, $(X, W), (\Omega, \mathcal{F}, P), \{\mathcal{F}_t\}_{t\geq 0}$ and $(\tilde{X}, \tilde{W}), (\tilde{\Omega}, \tilde{\mathcal{F}}, \tilde{P}), \{\tilde{\mathcal{F}}_t\}_{t\geq 0}$, and for every Borel set $A \subset \mathbb{B}(\mathbb{R}^d)$,

$$P(X_0 \in A) = \widetilde{P}(\widetilde{X}_0 \in A).$$

In other words, they have the same initial distribution.

Here the main focus is on Navier-Stokes and for a study of other well-established fluid models such as those given by Boussinesq equation and Magento-Hydrodynamic equation (MHD) we recommend [9,12,15]. In addition, more material on incompressible and compressible Navier-Stokes equations may be found in [17,30,31,33] and for a deeper study of stochastic analysis, we recommend [11,22,23,26]. As in the previous chapter, we denote vector $\vec{x}$ as x and let K be an arbitrary constant that may take different values from line to line. We also note that to keep the flow of the book, due to the heavy details of the articles discussed in this chapter, we discuss the main ideas and techniques and provide as many details on the steps as possible to prepare one to read and fully understand the papers.

8.1 Stochastic Navier Stokes Equation

We consider the stochastic Navier-Stokes equation with multiplicative noise given by,

$$\frac{\partial u^\varepsilon(t)}{\partial t} - \Delta u^\varepsilon(t) + (u^\varepsilon(t) \cdot \nabla)u^\varepsilon(t) + \nabla p(t,x) = f(t) + \sqrt{\varepsilon}\sigma(t, u^\varepsilon(t))\frac{dW(t)}{dt}, \tag{8.16}$$

where W is a Wiener process. As discussed in the introduction, for incompressible flow, we require $\operatorname{div}(u^\varepsilon(t)) = (\nabla \cdot u^\varepsilon)(t) = 0$. Here $u^\varepsilon(t)$ depends on both t and x, however, for better presentation, it is written as $u^\varepsilon(t)$.

Considering the open bounded domain, $D \subset \mathbb{R}^2$, and a Hilbert space, H, the Helmholtz-Leray projection, $P_H : L^2(D; \mathbb{R}^2) \to H$ is applied to each term in equation (8.16) above to transform it to a more compact form as follows,

$$du^\varepsilon(t) + Au^\varepsilon(t)dt + B(u^\varepsilon(t))dt = f(t)dt + \sqrt{\varepsilon}\sigma(t, u^\varepsilon(t))dW(t), \tag{8.17}$$

where

$$Au = -P_H\Delta u, \quad B(u) = B(u,u) = P_H((u \cdot \nabla)u), \quad P_H\nabla p = 0. \tag{8.18}$$

The operator A is referred to as the Stokes operator, which is positive-definite and self-adjoint and may be precisely defined as,

$$(Au, v) = \sum_{i,j=1}^{2} \int_D \partial_i u_j \, \partial_i v_j dx. \tag{8.19}$$

The above series implies that $(Au, u) = \sum_{i,j=1}^{2} \int_D |\partial_i u_j|^2 dx = \|\nabla u\|_{L^2}^2$. On the other hand, operator B is given by,

$$(B(u, v), w) = \sum_{i,j=1}^{2} \int_D u_i \, \partial_i v_j \, w_j dx, \tag{8.20}$$

which is often denoted as $(B(u, v), w) := b(u, v, w)$. Note that by the integration by parts, $(B(u, v), w) = -(B(u, w), v)$, implying, $b(u, v, w) = -b(u, w, v)$. Thus, if $v = w$, we have $b(u, v, v) = -b(u, v, v)$, that is $b(u, v, v) = 0$. Therefore, if the last two components of $b(u, v, w)$ are the same, then $b(u, v, w) = 0$.

Next we concentrate on the spaces used for Navier-Stokes equations. Let H and V be the closures of smooth, compactly supported functions in L^2 and H^1, respectively that are divergence-free. In other words, H is composed of all functions f in L^2 such that $\text{div} f = 0$ and function f is smooth and has compact support. The norm in H is given by,

$$|u| := \left(\int_D |u|^2 dx \right)^{1/2}, \tag{8.21}$$

and the norm in V is defined by,

$$\|u\| := \left(\int_D |\nabla u|^2 dx \right)^{1/2}. \tag{8.22}$$

In the study of the Navier-Stokes equations, notations $|\cdot|$ and $\|\cdot\|$ are typically used for the norms $\|\cdot\|_H$ and $\|\cdot\|_V$, respectively. Denoting the dual spaces of H and V as H' and V', respectively, we have the Gelfand triplet,

$$V \hookrightarrow H = H' \hookrightarrow V',$$

where the embeddings are continuous, dense and compact.

The article [35] by R. Wang, J. Zhai and T. Zhang is an excellent source for one to start his/her study of stochastic PDEs. It offers clear steps and uses techniques that are often applied in stochastic analysis. To introduce these techniques, we explain its first result, which is the Central limit theorem for stochastic Navier-Stokes equations. Contrary to large and moderate deviations, the Central limit theorem describes typical events. Recall that a sequence of independent, identically distributed random variables, $\{X_j\}_{j\geq 1}$ with $\mathbb{E}(X_j) = \mu$ and $Var(X_j) = \sigma^2$ satisfies the Central limit theorem if

$$\frac{S_n - n\mu}{\sigma\sqrt{n}} \xrightarrow{d} Y,$$

where Y has distribution, $N(0,1)$ and $S_n := \sum_{j=1}^{n} X_j$. For SPDEs, one needs to prove that the centered process divided by $\sqrt{\varepsilon}$,

$$Z^\varepsilon(t) = \frac{u^\varepsilon(t) - u^0(t)}{\sqrt{\varepsilon}}, \tag{8.23}$$

converges in distribution to a stochastic equation as $\varepsilon \to 0$. The authors in [35] use the well-posedness of the solutions to (8.16) in space $\mathcal{C}([0,T]; H) \cap L^2(0,T; V)$ provided by [9,32], where for $t \in [0,T]$,

$$\|u\|_{\mathcal{C}([0,T];H)\, \cap\, L^2(0,T;V)} := \sup_{0\leq t\leq T} |u(t)|^2 + \int_0^T \|u(s)\|^2 ds.$$

Then to achieve the Central limit theorem, they denote $Z^\varepsilon(t)$ as $V^\varepsilon(t)$ and letting $V^0(t)$ be given by,

$$dV^0(t) + \left(AV^0(t) + B(V^0(t), u^0(t)) + B(u^0(t), V^0(t))\right) dt = \sigma(t, u^0(t))dW(t), \tag{8.24}$$

they prove that,

$$\lim_{\varepsilon\to 0} \mathbb{E}\left(\left\|V^\varepsilon(t) - V^0(t)\right\|_{\mathcal{C}([0,T];H)\, \cap\, L^2(0,T;V)}\right) = 0. \tag{8.25}$$

As in [9,32], they consider the noise to be a Q-Wiener process and impose the following usual conditions on the noise coefficient and external force and use space $L_Q(H_0, H)$ defined in the introduction with norm (8.9).
i. The coefficient, $\sigma(t, u)$ takes values in $\mathcal{C}([0,T] \times V; L_Q(H_0; H))$ and for every $u, v \in V$,

$$\|\sigma(t, u)\|_{L_Q}^2 \leq K(1 + \|u\|^2) \quad \text{and} \quad \|\sigma(t, u) - \sigma(t, v)\|_{L_Q}^2 \leq K\|u - v\|^2. \tag{8.26}$$

ii. The external force, f takes values in $L^4([0,T]; V')$ implying that $\int_0^T \|f(s)\|_{V'}^4 ds < \infty$. They also use the following inequalities established in [32,33],

$$|b(u,v,w)| \leq 2\|u\|^{1/2} \cdot |u|^{1/2} \cdot \|v\|^{1/2} \cdot |v|^{1/2} \cdot \|w\|, \tag{8.27}$$

$$|b(u,u,v)| \leq \frac{1}{2}\|u\|^2 + c\|v\|_{L^4}^4 \cdot |u|^2, \tag{8.28}$$

$$|(B(u) - B(v), u - v)| \leq \frac{1}{2}\|u - v\|^2 + c|u - v|^2\|v\|_{L^4}^4. \tag{8.29}$$

By (8.17), $u^\varepsilon(t)$ and $u^0(t)$ are given by,

$$u^\varepsilon(t) + \int_0^t Au^\varepsilon(s)ds = u^\varepsilon(0) - \int_0^t B(u^\varepsilon(s))ds + \int_0^t f(s)ds$$

$$+ \sqrt{\varepsilon} \int_0^t \sigma(s, u^\varepsilon(s))dW(s),$$

and

$$u^0(t) + \int_0^t Au^0(s)ds = u^0(0) - \int_0^t B(u^0(s))ds + \int_0^t f(s)ds,$$

respectively. The authors prove in Proposition 3.1 of the paper that there exists an $\varepsilon_0 > 0$ such that for every $0 < \varepsilon \leq \varepsilon_0$,

$$\mathbb{E} \left\| u^\varepsilon(t) - u^0(t) \right\|_{\mathcal{C}([0,T];H) \cap L^2(0,T;V)} \leq \varepsilon K, \tag{8.30}$$

where K depends on $f(t)$ and T. Note that $V^\varepsilon(t)$ defined by (8.23) satisfies,

$$V^\varepsilon(t) + \int_0^t AV^\varepsilon(s)ds = V^\varepsilon(0) - \frac{1}{\sqrt{\varepsilon}} \int_0^t \Big(B(u^\varepsilon(s)) - B(u^0(s)) \Big) ds$$

$$+ \int_0^t \sigma(s, u^\varepsilon(s))dW(s).$$

Since both initial conditions, $u^\varepsilon(0)$ and $u^0(0)$ take the same value, then we have, $V^\varepsilon(0) = 0$. Also using (8.18), we may find,

$$\frac{1}{\sqrt{\varepsilon}} \Big(B(u^\varepsilon(s)) - B(u^0(s)) \Big) = \frac{1}{\sqrt{\varepsilon}} \Big(B(u^\varepsilon(s), u^\varepsilon(s)) - B(u^0(s), u^0(s)) \Big)$$

$$= \frac{1}{\sqrt{\varepsilon}} \Big(B(u^\varepsilon(s), u^\varepsilon(s)) - B(u^0(s), u^\varepsilon(s)) + B(u^0(s), u^\varepsilon(s)) - B(u^0(s), u^0(s)) \Big)$$

$$= B(V^\varepsilon(s), u^\varepsilon(s)) + B(u^0(s), V^\varepsilon(s)),$$

arriving at,

$$V^\varepsilon(t) + \int_0^t AV^\varepsilon(s)ds = -\int_0^t B(V^\varepsilon(s), u^\varepsilon(s))ds - \int_0^t B(u^0(s), V^\varepsilon(s))ds$$

$$+ \int_0^t \sigma(s, u^\varepsilon(s))dW(s). \tag{8.31}$$

Similar to the proof of (8.30), the authors prove in Lemma 3.2 of the paper that there exists an $\varepsilon_0 > 0$, such that,

$$\sup_{0\leq\varepsilon\leq\varepsilon_0} \mathbb{E}\left(\sup_{0\leq s\leq T} |V^\varepsilon(s)|^4 + 2\int_0^T |V^\varepsilon(s)|^2 \|V^\varepsilon(s)\|^2 ds \right) < K. \tag{8.32}$$

In addition, they note earlier in the paper that the following estimate holds for $V^0(t)$,

$$\mathbb{E}\left(\|V^0(t)\|_{\mathcal{C}([0,T];H) \cap L^2(0,T;V)} \right) < K. \tag{8.33}$$

In the paper, the authors apply the widely used inequality in stochastic analysis called the Burkholder-Davis-Gundy inequality given as follows. For an $\mathcal{F}_t$-continuous martingale, M_t with increasing process, $< M >_t$, for every $p \geq 1$ and stopping time, τ,

$$\mathbb{E}\left(\left| \sup_{0\leq t\leq\tau} |M_t| \right|^p \right) \leq K\mathbb{E}\left(\langle M \rangle_\tau^{\frac{p}{2}} \right). \tag{8.34}$$

In above, $\langle M \rangle_t$ denotes the quadratic variation of M_t, which in the case of a stochastic integral, $\int_0^t \sigma(s)dB(s)$, is $\int_0^t \sigma(s)^2 ds$. More precisely, since stochastic integrals, $\int_0^t \sigma(s)dB(s)$ are martingales if $\int_0^T \mathbb{E}\left(\sigma(s)^2\right)ds < \infty$, then in this case, the Burkholder-Davis-Gundy inequality becomes,

$$\mathbb{E}\left(\sup_{0\leq t\leq T} \left| \int_0^t \sigma(s)dB(s) \right|^p \right) \leq K\mathbb{E}\left(\int_0^T |\sigma(s)|^2 ds \right)^{\frac{p}{2}}.$$

Theorem 8.1 *(Theorem 3.2 in [35]) There exists $\varepsilon_0 > 0$ such that the family, $\{u^\varepsilon(\cdot)\}_{\varepsilon_0\leq\varepsilon}$ of solutions to (8.17) satisfies the Central limit theorem in $\mathcal{C}([0,T]; H) \cap L^2(0,T; V)$.*

Proof. To prove (8.25), we find using (8.31) and (8.24),

$$\left(V^\varepsilon(t) - V^0(t)\right) + \int_0^t A\left(V^\varepsilon(s) - V^0(s)\right) ds$$

$$= -\int_0^t \left(B(V^\varepsilon(s), u^\varepsilon(s)) - B(V^0(s), u^0(s))\right) ds$$

$$-\int_0^t B\left(u^0(s), V^\varepsilon(s) - V^0(s)\right) ds + \int_0^t \left(\sigma(s, u^\varepsilon(s)) - \sigma(s, u^0(s))\right) dW(s).$$

Applying Itô's formula (8.7) to $V^\varepsilon(t) - V^0(t)$ as in (8.10), we obtain,

$$\left|V^\varepsilon(t) - V^0(t)\right|^2 + 2\int_0^t \left(A\left(V^\varepsilon(s) - V^0(s)\right), V^\varepsilon(s) - V^0(s)\right) ds$$

$$= -2\int_0^t \left(B(V^\varepsilon(s), u^\varepsilon(s)) - B(V^0(s), u^0(s)), V^\varepsilon(s) - V^0(s)\right) ds$$

$$- 2\int_0^t \left(B\left(u^0(s), V^\varepsilon(s) - V^0(s)\right), V^\varepsilon(s) - V^0(s)\right) ds$$

$$+ 2\int_0^t \left(\left(\sigma(s, u^\varepsilon(s)) - \sigma(s, u^0(s))\right) dW(s), V^\varepsilon(s) - V^0(s)\right) ds$$

$$+ \int_0^t \left\|\sigma(s, u^\varepsilon(s)) - \sigma(s, u^0(s))\right\|_{L_Q}^2 ds. \qquad (8.35)$$

Recall that by (8.19), $(Au, u) = \|\nabla u\|_{L^2}^2 = \|u\|_V^2$ and $(B(u, v), w)$, written as $b(u, v, w)$, satisfies, $b(u, v, v) = 0$. Thus, the second term on the right hand side drops. To ensure that the left hand side is bounded, we let $\tau_N := \inf\left\{0 \le t \le T : \|V^\varepsilon(t) - V^0(t)\|_{C([0,T];H) \cap L^2(0,T;V)} > N\right\}$ be a stopping time and make the integrals' bounds in (8.35) from zero to $\min\{t, \tau_N\}$, denoted by $t \wedge \tau_N$. Taking the absolute value of the right hand side, then the supremum up to time $t \wedge \tau_N$ and afterwards the expectation of both sides, we have,

$$\mathbb{E}\left(\sup_{0 \le s \le t \wedge \tau_N} \left|V^\varepsilon(s) - V^0(s)\right|^2 + \sup_{0 \le s \le t \wedge \tau_N} \int_0^s \left\|V^\varepsilon(\ell) - V^0(\ell)\right\|^2 d\ell\right)$$

$$\le 2\mathbb{E}\int_0^{t \wedge \tau_N} \left|\left(B\left(V^\varepsilon(s), u^\varepsilon(s) - u^0(s)\right), V^\varepsilon(s) - V^0(s)\right)\right| ds$$

$$+ 2\,\mathbb{E} \int_0^{t\wedge\tau_N} \left|\left(B\left(V^\varepsilon(s) - V^0(s), u^0(s)\right), V^\varepsilon(s) - V^0(s)\right)\right| ds$$

$$+ 2\,\mathbb{E}\left(\sup_{0\le s\le t\wedge\tau_N} \left|\int_0^s \left(\left(\sigma(\ell, u^\varepsilon(\ell)) - \sigma(\ell, u^0(\ell))\right) dW(\ell), V^\varepsilon(\ell) - V^0(\ell)\right)\right|\right)$$

$$+ \mathbb{E}\int_0^{t\wedge\tau_N} \left\|\sigma(s, u^\varepsilon(s)) - \sigma(s, u^0(s))\right\|^2_{L_Q} ds$$

$$= I_1 + I_2 + I_3 + I_4,$$

where in the first term on the right hand side of (8.35), we added and sub-
tracted term, $B(V^\varepsilon(s), u^0(s))$. Note that the first term may be bounded by,

$$I_1 \le 2\sqrt{\varepsilon}\,\mathbb{E}\int_0^{t\wedge\tau_N} \left|\left(B\left(V^\varepsilon(s), V^\varepsilon(s)\right), V^\varepsilon(s)\right)\right| ds$$

$$+ 2\sqrt{\varepsilon}\,\mathbb{E}\int_0^{t\wedge\tau_N} \left|\left(B\left(V^\varepsilon(s), V^\varepsilon(s)\right), V^0(s)\right)\right| ds,$$

in which the first term may be dropped. Applying inequality (8.27) and then
Youngs inequality on the second term in the bound of I_1 and using inequality
(8.28) to bound I_2, we have by (8.32) and (8.33),

$$I_1 + I_2 \le 2\sqrt{\varepsilon}\,\mathbb{E}\int_0^{t\wedge\tau_N} |V^\varepsilon(s)|^2 \left\|V^\varepsilon(s)\right\|^2 ds + 2\sqrt{\varepsilon}\,\mathbb{E}\int_0^{t\wedge\tau_N} \left\|V^0(s)\right\|^2 ds$$

$$+ \mathbb{E}\int_0^{t\wedge\tau_N} \left\|V^\varepsilon(s) - V^0(s)\right\|^2 ds + 2K\mathbb{E}\int_0^{t\wedge\tau_N} \left|V^\varepsilon(s) - V^0(s)\right|^2 \|u^0(s)\|^4_{L^4} ds$$

$$\le \mathbb{E}\int_0^{t\wedge\tau_N} \left\|V^\varepsilon(s) - V^0(s)\right\|^2 ds + K\mathbb{E}\int_0^{t\wedge\tau_N} \left|V^\varepsilon(s) - V^0(s)\right|^2 \|u^0(s)\|^4_{L^4} ds.$$

Furthermore, by conditions in (8.26) on the noise coefficient and noting
inequality (8.30),

$$I_4 \le K\mathbb{E}\int_0^{t\wedge\tau_N} \left\|u^\varepsilon(s) - u^0(s)\right\|^2 ds \le K\varepsilon.$$

Now we apply the Burkholder-Davis-Gundy inequality on the third term to
obtain,

$$I_3 \le K\mathbb{E}\left[\int_0^{t\wedge\tau_N} \left\|\sigma(s, u^\varepsilon(s)) - \sigma(s, u^0(s))\right\|^2_{L_Q} \left|V^\varepsilon(s) - V^0(s)\right|^2 ds\right]^{\frac{1}{2}},$$

which may be simplified using (8.26) and then Young's inequality and (8.30)
as follows,

$$\leq K \mathbb{E} \left(\sup_{0 \leq s \leq t \wedge \tau_N} \left| V^\varepsilon(s) - V^0(s) \right|^2 \int_0^{t \wedge \tau_N} \left\| u^\varepsilon(s) - u^0(s) \right\|^2 ds \right)^{\frac{1}{2}}$$

$$= K \mathbb{E} \left[\left(\sup_{0 \leq s \leq t \wedge \tau_N} \left| V^\varepsilon(s) - V^0(s) \right| \right) \left(\int_0^{t \wedge \tau_N} \left\| u^\varepsilon(s) - u^0(s) \right\|^2 ds \right)^{\frac{1}{2}} \right]$$

$$\leq \frac{1}{2} \mathbb{E} \left(\sup_{0 \leq s \leq t \wedge \tau_N} \left| V^\varepsilon(s) - V^0(s) \right|^2 \right) + \frac{K^2}{2} \mathbb{E} \left(\int_0^{t \wedge \tau_N} \left\| u^\varepsilon(s) - u^0(s) \right\|^2 ds \right)$$

$$\leq \frac{1}{2} \mathbb{E} \left(\sup_{0 \leq s \leq t \wedge \tau_N} \left| V^\varepsilon(s) - V^0(s) \right|^2 \right) + K \varepsilon.$$

Hence, we obtain,

$$\mathbb{E} \left(\sup_{0 \leq s \leq t \wedge \tau_N} \left| V^\varepsilon(s) - V^0(s) \right|^2 + 2 \sup_{0 \leq s \leq t \wedge \tau_N} \int_0^s \left\| V^\varepsilon(\ell) - V^0(\ell) \right\|^2 d\ell \right)$$

$$\leq K \varepsilon + \mathbb{E} \left(\int_0^{t \wedge \tau_N} \left\| V^\varepsilon(s) - V^0(s) \right\|^2 ds \right)$$

$$+ K \mathbb{E} \left(\int_0^{t \wedge \tau_N} \sup_{0 \leq \ell \leq s} \left| V^\varepsilon(\ell) - V^0(\ell) \right|^2 \| u^0(s) \|_{L^4}^4 ds \right),$$

which by grouping $\mathbb{E} \int_0^{t \wedge \tau_N} \| V^\varepsilon(s) - V^0(s) \|^2 ds$ on both sides becomes,

$$\mathbb{E} \left(\sup_{0 \leq s \leq t \wedge \tau_N} \left| V^\varepsilon(s) - V^0(s) \right|^2 + \int_0^{t \wedge \tau_N} \left\| V^\varepsilon(s) - V^0(s) \right\|^2 ds \right)$$

$$\leq K \varepsilon + K \mathbb{E} \left(\int_0^{t \wedge \tau_N} \sup_{0 \leq \ell \leq s} \left| V^\varepsilon(\ell) - V^0(\ell) \right|^2 \| u^0(s) \|_{L^4}^4 ds \right).$$

Now the Gronwall's inequality (see the Appendix of Chapter 6) may be applied to $f(t) = \mathbb{E} \sup_{0 \leq s \leq t \wedge \tau_N} |V^\varepsilon(s) - V^0(s)|^2$ to attain,

$$\mathbb{E} \left(\sup_{0 \leq s \leq t \wedge \tau_N} \left| V^\varepsilon(s) - V^0(s) \right|^2 + \int_0^{t \wedge \tau_N} \left\| V^\varepsilon(s) - V^0(s) \right\|^2 ds \right)$$

$$\leq K \varepsilon \exp \left(K \int_0^T \| u^0(s) \|_{L^4}^4 ds \right).$$

Since the above inequality holds for any $N > 0$, then we set $N \to \infty$ and $\varepsilon \to 0$ to achieve (8.25). □

We now explain a well-posedness result by applying the Galerkin approximation method. In [15], J. Duan and A. Millet consider the following equation,

$$\phi_h^\varepsilon(t) = - \int_0^t A\phi_h^\varepsilon(s)ds - \int_0^t B(\phi_h^\varepsilon(s))ds - \int_0^t R\phi_h^\varepsilon(s)ds$$

$$+ \int_0^t \widetilde{\sigma}(\phi_h^\varepsilon(s))h(s)ds + \sqrt{\varepsilon} \int_0^t \sigma(\phi_h^\varepsilon(s))dW(s), \qquad (8.36)$$

which is the stochastic Navier-Stokes equation (8.17) without the external force and with two added terms. As we discuss in Sections 8.2 and 8.3, this equation is used in the theory of large deviations. For similar well-posedness results we refer to Theorem 3.1 in [1] and Theorem 2.4 in [9]. Recall that the Galerkin approximation method is applied to infinite-dimensional PDEs to prove the well-posedness of solutions. The method first applies a projection to obtain the finite-dimensional analogue of the equation, for which the existence and uniqueness of solutions are less difficult to attain. Then based on the uniform bounds on the finite dimensional equation, weak convergence limits are obtained that can be used to pass each term to its limit. After proving that the limit has a unique solution and that each term in the limit matches the terms of the original equation, the existence and uniqueness of the original equation is achieved. For examples of proving the existence and uniqueness of solutions by the Galerkin approximations for deterministic PDEs, we recommend [36], in which the method is applied to various equations.

For the above steps, we need the following theorems. Recall that a sequence, $(u_n)_n$ in X converges weakly if for all $v \in X'$, $\langle u_n, v \rangle \to \langle u, v \rangle$ as $n \to \infty$ and a sequence $(v_n)_n$ in X' converges weakly star to $v \in X'$ if for every $u \in X$, $\langle v_n, u \rangle \to \langle v, u \rangle$ strongly as $n \to \infty$.

Theorem 8.2 (Proposition 21.23 (b, g, i, j, k) in [36])   *Let $(u_n)_n$ be a sequence in the Banach space, X over $\mathbb{R}$ or $\mathbb{C}$ then,*

    i. *If X is finite dimensional, then weak convergence implies strong convergence.*

    ii. *If $(u_n)_n$ is bounded in X and there exist $u \in X$ and a dense set $D \subset X'$, such that for all $f \in D$, $\langle f, u_n \rangle \to \langle f, u \rangle$ strongly as $n \to \infty$, then $u_n \to u$ weakly as $n \to \infty$.*

    iii. *A bounded sequence in a reflexive Banach space has a weakly convergent subsequence. In addition, if each weakly convergent subsequence has the same limit, then the original bounded sequence converges weakly to that limit.*

    iv. *If $u_n \to u$ weakly in X and $f_n \to f$ strongly in X', then $\langle f_n, u_n \rangle \to \langle f, u \rangle$ strongly.*

v. If X is reflexive and $u_n \to u$ strongly in X and $f_n \to f$ weakly in X', then $\langle f_n, u_n \rangle \to \langle f, u \rangle$ strongly.

Theorem 8.3 (Proposition 21.26(f) in [36]) *In a reflexive Banach space, X, over $\mathbb{R}$ or $\mathbb{C}$, weak star convergence implies weak convergence. Namely, if $(v_n)_n$ is a sequence in X' and $v_n \to v$ weakly star, then $v_n \to v$ weakly as $n \to \infty$.*

Theorem 8.4 *(Theorem 3.1 in [15]) There exists a unique weak in PDE and strong in probability solution to equation (8.36).*

Proof. To obtain the finite dimensional analogue of the original equation, let $\{\varphi_n\}_{n \geq 1}$ be an orthonormal basis of space H, such that each φ_n is in the domain of the Stokes operator, A. That is, $\varphi_n \in \text{Dom}(A)$. Then letting $H_n = \text{span}\{\varphi_1, ..., \varphi_n\}$, the orthogonal projection, $P_n : H \to H_n$ from H to H_n is obtained and applying this projection to each term of the original equation, gives us the finite dimensional equation, $\phi^\varepsilon_{n,h}(t)$ as follows,

$$\phi^\varepsilon_{n,h}(t) = -\int_0^t A\phi^\varepsilon_{n,h}(s)ds - \int_0^t B(\phi^\varepsilon_{n,h}(s))ds - \int_0^t R(\phi^\varepsilon_{n,h}(s))ds$$
$$+ \int_0^t \tilde{\sigma}_n(\phi^\varepsilon_{n,h}(s))h(s)ds + \sqrt{\varepsilon} \int_0^t \sigma_n(\phi^\varepsilon_{n,h}(s))dW_n(s), \qquad (8.37)$$

where $W_n = P_n W$, $\sigma_n = P_n \sigma$ and $\tilde{\sigma}_n = P_n \tilde{\sigma}$. By the conditions assumed and properties of the coefficients in (8.37), we have that each is globally or at least locally Lipschitz continuous, therefore, (8.37) has a unique maximal solution. For a definition of maximal solution we refer the reader to the introduction of this chapter and for a result obtaining the unique maximal solution by requiring the local Lipschitz continuity of the coefficients see for example Theorem 3.7 of [19]. The authors next apply the Itô's formula along with a special case of Gronwall's inequality, which they establish in the paper's Lemma 3.9, to obtain,

$$\sup_n \mathbb{E}\left(\sup_{0 \leq t \leq T} \left|\phi^\varepsilon_{n,h}(t)\right|^{2p} + \int_0^T \left\|\phi^\varepsilon_{n,h}(s)\right\|^2 \left|\phi^\varepsilon_{n,h}(s)\right|^{2(p-1)} ds\right) \leq K\left(\mathbb{E}|\xi|^{2p} + 1\right),$$
$$(8.38)$$

where it is assumed that $\phi^\varepsilon_{n,h}(0) = \xi \in L^{2p}(\Omega, H)$. Here we concentrate on the well-posedness result and note that similar estimates as those carried out in this paper, are explained in more detail in Section 8.3.

To extend the maximal solution of (8.37) to a global solution, let $\tau_N := \inf\left\{t : |\phi^\varepsilon_{n,h}(t)| \geq N\right\} \wedge T$, then (8.37) holds on time interval, $[0, \tau_N \wedge t]$. Since $\phi^\varepsilon_{n,h}(t)$ is a maximal solution, then there exists a stopping time, $\tau_\infty \leq T$, such that $\tau_N \nearrow \tau_\infty$ a.s. as $N \to \infty$ and $\sup_{0 \le s \le t_\infty} \left| \phi^\varepsilon_{n,h}(s) \right| \to \infty$, which is a contradiction to estimate (8.38). Thus, $\tau_{n,h} = T$ and the solution to (8.37) is global.

By (8.38), $\phi^\varepsilon_{n,h}(t)$ is bounded and we may apply Theorem 8.2(*iii*) to obtain a convergent subsequence, which still will be denoted as $\{\phi^\varepsilon_{n,h}\}_n$. We let the weak and weak star convergence be denoted by $\rightharpoonup$ and $\overset{*}{\rightharpoonup}$, respectively. Setting $p = 1$ in estimate (8.38), we find,

$$\phi^\varepsilon_{n,h}(T) \rightharpoonup \tilde{\phi}^\varepsilon_h(T) \text{ in } L^2(\Omega, H) \text{ and } \phi^\varepsilon_{n,h} \rightharpoonup \phi^\varepsilon_h \text{ in } L^2(\Omega, V),$$

for a $\tilde{\phi}^\varepsilon_h(T) \in L^2(\Omega, H)$. Also with $p = 2$ we have by the first term on the left hand side of (8.38),

$$\phi^\varepsilon_{n,h} \overset{*}{\rightharpoonup} \phi^\varepsilon_h \text{ in } L^4(\Omega, L^\infty([0, T]; H)).$$

Using estimates and conditions on each of the coefficients in (8.37), the expectation of each term may be bounded by a constant multiple of an expression involving $\phi^\varepsilon_{n,h}$ or $u^\varepsilon_{n,h}$ and thus is uniformly bounded and the following convergence limits are attained in $L^2(\Omega, V')$.

$$A\phi^\varepsilon_{n,h} \rightharpoonup A^\varepsilon_h \text{ and } B\phi^\varepsilon_{n,h} \rightharpoonup B^\varepsilon_h \text{ and } R\phi^\varepsilon_{n,h} \rightharpoonup R^\varepsilon_h,$$

and for $S^\varepsilon_h, \tilde{S}^\varepsilon_h \in L^2(\Omega, L_Q)$, the authors obtain by some estimates,

$$\sigma_n(\phi^\varepsilon_{n,h})P_n \rightharpoonup S^\varepsilon_h \text{ in } L^2(\Omega, L_Q), \text{ and } \tilde{\sigma}_n(\phi^\varepsilon_{n,h})h \rightharpoonup \tilde{S}^\varepsilon_h \text{ in } L^{\frac{4}{3}}(\Omega, H).$$

In Proposition 3.11 of the paper, they also find,

$$\sup_n \mathbb{E} \int_0^T |\phi^\varepsilon_{n,h}(s)|^4_{L^4} ds \le K(1 + \mathbb{E}|\xi|^4),$$

from which the following convergence may be deduced,

$$\phi^\varepsilon_{n,h} \rightharpoonup \phi^\varepsilon_h \text{ in } L^4(\Omega, L^4(D)^3).$$

Now with the orthonormal basis, $\{\varphi_n\}_{n \ge 1}$ used earlier for the projection to the finite dimensional equation, let $g_j(t) = f(t)\varphi_j$, where for $\delta > 0$, f is in $H^1(-\delta, T+\delta)$ satisfying, $\|f\|_\infty = 1$ and $f(0) = 1$. With the test function, $g_j(t)$, we obtain,

$$\left(\phi_{n,h}^{\varepsilon}(T), g_j(T) \right) = (\xi, \varphi_j) + \int_0^T \left(\phi_{n,h}^{\varepsilon}(s), \varphi_j \right) f'(s) ds - \int_0^T \left(A\phi_{n,h}^{\varepsilon}, g_j(s) \right) ds \tag{8.39}$$

$$- \int_0^T \left(B\phi_{n,h}^{\varepsilon}, g_j(s) \right) ds - \int_0^T \left(R\phi_{n,h}^{\varepsilon}, g_j(s) \right) ds$$

$$+ \sqrt{\varepsilon} \int_0^T \left(\sigma_n(\phi_{n,h}^{\varepsilon}(s)) dW_n(s), g_j(s) \right)$$

$$+ \int_0^T \left(\tilde{\sigma}_n(\phi_{n,h}^{\varepsilon}(s)) h(s), g_j(s) \right) ds,$$

from which by applying the limits found above and Theorems 8.2(*iv*) and 8.3, we have the strong convergence of (8.39) to,

$$(\tilde{\phi}_h^{\varepsilon}(T), \varphi_j) f(T) = (\xi, \varphi_j) + \int_0^T (\phi_h^{\varepsilon}(s), \varphi_j) f'(s) ds - \int_0^T (A_h^{\varepsilon}(s), g_j(s)) ds$$

$$- \int_0^T (B_h^{\varepsilon}(s), g_j(s)) ds - \int_0^T (R_h^{\varepsilon}(s), g_j(s)) ds$$

$$+ \sqrt{\varepsilon} \int_0^T (S_h^{\varepsilon}(s) dW(s), g_j(s)) + \int_0^T \left(\tilde{S}_h^{\varepsilon}(s), g_j(s) \right) ds.$$

Note that by definition, $g_j(0) = \varphi_j$. Now choosing, $f = 1_{(-\delta, T+\delta)}$, we arrive at,

$$\phi_h^{\varepsilon}(T) = \xi - \int_0^T A_h^{\varepsilon}(s) ds - \int_0^T B_h^{\varepsilon}(s) ds - \int_0^T R_h^{\varepsilon}(s) ds$$

$$+ \sqrt{\varepsilon} \int_0^T S_h^{\varepsilon}(s) dW(s) + \int_0^T \tilde{S}_h^{\varepsilon}(s) ds. \tag{8.40}$$

Next the authors verify that (8.40) is indeed the solution to (8.36) with control h by proving that a.s.,

$$A_h^{\varepsilon}(t) = A(\phi_h^{\varepsilon}(t)), \quad B_h^{\varepsilon}(t) = B(\phi_h^{\varepsilon}(t)), \quad R_h^{\varepsilon}(t) = R(\phi_h^{\varepsilon}(t)),$$

$$S_h^{\varepsilon}(t) = \sigma(\phi_h^{\varepsilon}(t)), \quad \tilde{S}_h^{\varepsilon}(t) = \tilde{\sigma}(\phi_h^{\varepsilon}(t)) h(t).$$

For this verification, they consider the test function, $\tilde{\psi} \in L^{\infty}(\Omega_T, H_m)$ for a fixed m and show that,

$$\mathbb{E} \int_0^T \left(A_h^\varepsilon(s) - A(\phi_h^\varepsilon(s)), \widetilde{\psi} \right) ds + \mathbb{E} \int_0^T \left(B_h^\varepsilon(s) - B(\phi_h^\varepsilon(s)), \widetilde{\psi} \right) ds$$

$$+ \mathbb{E} \int_0^T \left(R_h^\varepsilon(s) - R(\phi_h^\varepsilon(s)), \widetilde{\psi} \right) ds + \mathbb{E} \int_0^T \left(\widetilde{S}_h^\varepsilon(s) - \widetilde{\sigma}(\phi_h^\varepsilon(s)), \widetilde{\psi} \right) ds = 0.$$

$$(8.41)$$

We leave the details on deriving (8.41) and obtaining the uniqueness of solutions to (8.36) to the interested reader, since the proofs require many detailed steps and estimates. □

8.2 Large Deviations

Large deviations is a branch of probability theory that studies events that "largely deviate" from typical events. That is, events that one does not expect to occur and are too extreme that are often avoided. In the language of probability, these events have probability going to zero exponentially fast and are said to satisfy the large deviation principle, which is formally given as follows. A family of random variables, $\{u^\varepsilon\}_{\varepsilon>0}$, taking values in space, $\mathcal{E}$, satisfies the large deviations principle if for every open set, $G \subset \mathcal{E}$,

$$- \inf_{x \in G} I(x) \le \liminf_{\varepsilon \to 0} \varepsilon \log P(u^\varepsilon \in G),$$

and for every closed set, $C \subset \mathcal{E}$,

$$\limsup_{\varepsilon \to 0} \varepsilon \log P(u^\varepsilon \in C) \le - \inf_{x \in C} I(x).$$

The two inequalities are referred to as large deviations lowerbound and upperbound, respectively. The function, $I : \mathcal{E} \to [0, \infty]$ in the bounds is called the rate function and as can be observed from the above, it gives the rate of exponential decay of the probability of the rare event in occurring. In the study of large deviations, researchers typically let space $\mathcal{E}$ be a Polish space, which means that it is a complete, separable, metric space. Choosing such spaces enables one to apply many of the results in the large deviation theory, the most useful of which is the equivalence of Laplace principle with large deviations. A family of random variables, $\{u^\varepsilon(\cdot)\}_{\varepsilon>0}$ taking values in $\mathcal{E}$, satisfies the Laplace principle on $\mathcal{E}$ if for every bounded continuous functions, h,

$$- \inf_{x \in \mathcal{E}} \{h(x) + I(x)\} \leq \liminf_{\varepsilon \to 0} \varepsilon \log \mathbb{E} \left(e^{-\frac{h(u^\varepsilon(t))}{\varepsilon}} \right),$$

and

$$\limsup_{\varepsilon \to 0} \varepsilon \log \mathbb{E} \left(e^{-\frac{h(u^\varepsilon(t))}{\varepsilon}} \right) \leq - \inf_{x \in \mathcal{E}} \{h(x) + I(x)\}.$$

For the proof of the equivalence of Laplace principle with large deviations principle in the case of Polish spaces and more general background on large deviations, we refer the reader to [5,14,16]. Large deviations has proven to be useful in many applied fields by providing the rate of decay probability of the process exceeding a given threshold value. We have included examples of such applications in the Appendix.

In the setting of stochastic differential equations, to prove the small-noise large deviations, one multiplies the noise term by $\sqrt{\varepsilon}$, where $\varepsilon > 0$. For example, in the case of stochastic Navier-Stokes equation, we let u depend on $\varepsilon > 0$ and consider equation (8.17). The goal is to the study of the behavior of the process as $\varepsilon \to 0$, which translates to the study of the rare event of setting the noise in the process to vanish.

Another closely related topic in large deviations is the study of moderate deviations, which studies events that are rare and atypical, but not as rare as the events described by large deviations. Thus, as the name indicates, they "moderately deviate" from the mean. Letting $u^0(t)$ be the process obtained by setting $\varepsilon = 0$, to prove the moderate deviation principle, one has to prove the large deviation principle for the centered process: $u^\varepsilon(t) - u^0(t)$ multiplied by a rate that converges to zero at a slower rate than that of large deviations. If $u^\varepsilon(t)$ denotes the solution of a stochastic differential or partial differential equation, then to achieve the moderate deviation principle for $\{u^\varepsilon(\cdot)\}_{\varepsilon>0}$, one proves the large deviation principle for the process,

$$v^\varepsilon(t) = \frac{a(\varepsilon)}{\sqrt{\varepsilon}} (u^\varepsilon(t) - u^0(t)),$$

where $a(\varepsilon)$ is a function of $\varepsilon > 0$ that satisfies,

$$a(\varepsilon) > 0 \quad \text{and} \quad \frac{a(\varepsilon)}{\sqrt{\varepsilon}} \to \infty \quad \text{as} \quad \varepsilon \to 0. \tag{8.42}$$

Observe that since the noise term in $u^\varepsilon(t)$ is multiplied by $\sqrt{\varepsilon}$, then the multiple of the noise term in $v^\varepsilon(t)$ becomes $a(\varepsilon)$. Also, the second condition in (8.42) guarantees that $a(\varepsilon)$ converges to zero at a slower rate than $\sqrt{\varepsilon}$, which is the multiple of the noise term in large deviations. Hence the rate of decay of $a(\varepsilon)$ is moderate compared to $\sqrt{\varepsilon}$.

Since it is often difficult to obtain the inequalities required for large and moderate deviations directly, some techniques in the literature have provided researchers alternative routes to establish the theory for a given process. Two of the most widely applied techniques are the Azencott method and the weak convergence approach. In this section, we will discuss one article for each method. More precisely, a result on large deviations for a class of stochastic Volterra equations is described based on the Azencott method and large deviations for stochastic Boussinesq equation is discussed for the weak convergence approach.

The Azencott method for proving the large deviation principle for a process is defined as follows. The process, $\{X^\varepsilon\}_{\varepsilon>0}$ taking values in the Polish space, $(\mathcal{E}_1, d_1)$ satisfies the large deviation principle if one can find another process, $\{Y^\varepsilon\}_{\varepsilon>0}$ taking values in the Polish space, $(\mathcal{E}_2, d_2)$ that is known to satisfy the large deviation principle with rate function, $I : \mathcal{E}_2 \to [0, \infty]$ and that the following inequality holds. For all positive constants, R, ρ, a, there exists $\alpha > 0$ and $\varepsilon_0 > 0$ such that for every $\varepsilon \leq \varepsilon_0$ and $f \in \mathcal{E}_2$ with $I(f) \leq a$,

$$P\left(d_1\left(X^\varepsilon, \Phi(f)\right) \geq \rho, \ d_2\left(Y^\varepsilon, f\right) < \alpha\right) \leq \exp\left(-\frac{R}{\varepsilon^2}\right), \qquad (8.43)$$

where $\Phi : \{I < \infty\} \to \mathcal{E}_1$ is continuous on compact sets, $\{I \leq a\}$ for $a > 0$ with respect to d_2. Note that notation, $\exp(x) := e^x$ is used for better presentation. The above inequality is referred to as the Freidlin-Wentzell inequality and its derivation is the key part of the proof of large deviations by Azencott method.

In the context of stochastic differential equations, the map, Φ is $h \to S(h)$ where $S(h)$ is the unique solution to the original stochastic equation's controlled equation, also called the skeleton equation. The controlled equation is obtained by replacing the Wiener process in the noise term in the original equation with a function, h in space L^2.

In [25], D. Nualart and C. Rovira establish the large deviations principle for the following class of SPDEs,

$$X_t^\varepsilon = x_0 + \int_0^t b(t, s, X_s^\varepsilon)ds + \sum_{j=1}^k \int_0^t \varepsilon\sigma_j(t, s, X_s^\varepsilon)dW_s^j, \qquad (8.44)$$

where $\{W^j(t) : 1 \leq j \leq k\}$ is a Wiener process. The process, $\{X^\varepsilon\}_\varepsilon$, depending on $0 < \varepsilon \leq 1$, takes values in the Polish space, $\mathcal{C}([0, T]; \mathbb{R}^d)$, which is the space of continuous functions, $f : [0, T] \to \mathbb{R}^d$. The above equation is called a Volterra equation since the integrand as well as the integral depend on t. The controlled PDE of (8.44) is given by,

$$S(h) = x_0 + \int_0^t b(t, s, S(h))ds + \sum_{j=1}^k \int_0^t \sigma_j(t, s, S(h))\dot{h}_s^j ds, \qquad (8.45)$$

where h is in the Cameron-Martin space, $\mathcal{H}$ defined as,

$$\mathcal{H} = \left\{ h \in \mathcal{C}([0, T]; \mathbb{R}^d) : \int_0^T |\dot{h}_s|^2 ds < \infty \right\}, \qquad (8.46)$$

with $\dot{f}_s$ denoting the derivative of f with respect to time. Thus, the Wiener process, W was replaced with $\dot{h}_s^j$, which is in L^2 and $\varepsilon > 0$ is taken out of the equation.

Recall that a function, $f(x)$ is α-Hölder continuous if there exists, $0 < \alpha \leq 1$ such that,

$$|f(x) - f(y)| \leq K|x - y|^\alpha, \qquad (8.47)$$

where in the case of $\alpha = 1$ yields the Lipschitz continuity. Also the linear growth condition is

$$|f(x)| \leq K(1 + |x|). \qquad (8.48)$$

Lipschitz continuity and linear growth condition are the most commonly imposed assumptions on the coefficients of SPDEs, since classical theorems may be used to obtain the well-posedness of solutions. In [25] the authors assume that the coefficients b and σ_j are bounded, measurable, Lipschitz continuous in x, and α-Hölder continuous in t where $0 < \alpha \leq 1$. They also assume the inequality,

$$\sum_{j=1}^k \left| \sigma_j(t, s, x) - \sigma_j(r, s, x) - \sigma_j(t, s, y) + \sigma_j(r, s, y) \right| \leq K|t - r|^\gamma |x - y|.$$

For better presentation, the authors drop the sums in equations (8.44) and (8.45) and prove the results for any arbitrary $j \in [1, k]$. Note that the Freidlin-Wentzell inequality in their case is given as follows. For any $g \in \mathcal{H}, R$ and $\rho > 0$, there exists $\eta > 0$, and $\varepsilon_0 > 0$ such that for all $0 < \varepsilon \leq \varepsilon_0$,

$$P\left(\|X^\varepsilon - S(g)\|_\infty > \rho, \ \|\varepsilon W - g\|_\infty < \eta \right) \leq \exp\left(-\frac{R}{\varepsilon^2} \right). \qquad (8.49)$$

Since X^ε takes values in $\mathcal{C}([0, T]; \mathbb{R}^d)$, then the uniform norm is the proper norm used for the difference, $X^\varepsilon - S(g)$.

To obtain the exponential bound required by the Friedlin-Wentzell inequality, many authors apply the following theorem.

Theorem 8.5 *(Theorem 3.3 in [7]) If $X_t = \int_0^t \sigma_s dW_s$ is an H-valued integral for a Hilbert space H and $\int_0^T \|\sigma\|_{L_Q} ds < \infty$, then for any $r > 0$ and $t \in [0, T]$,*

$$P\left(\sup_{0 \le s \le t} |X_s| \ge r\right) \le \exp\left(-\frac{r^2}{K}\right).$$

Since the integral in [25] is a Volterra integral, then the above theorem cannot be applied and the authors instead prove the following theorem to use for their result.

Theorem 8.6 *(Theorem 2 in [25]) If $Z : [0, T] \times [0, T] \times \Omega \to \mathbb{R}^d \times \mathbb{R}^k$, is a measurable process on $\mathbb{B}([0, T]) \otimes \mathbb{B}([0, T]) \otimes \mathcal{F}$ that satisfies,*

i. $Z(t, s) = 0$ if $s > t$,

ii. $Z(t, s)$ is $\mathcal{F}_s$-adapted,

iii. *for all $t, r \in [0, T]$, there exists $\alpha \in (0, 2]$ and a positive random variable, ξ such that,*

$$\int_0^{\min(r,t)} |Z(t, s) - Z(r, s)|^2 ds \le \xi |t - r|^\alpha,$$

then for all $0 < \beta \le \min(1, \alpha)$, nonnegative constants, L, C, K_1, K_2, K_3 such that $\max\{K_1, K_2(1 + T)T^{\beta/2}\} \le L(T^{\alpha-\beta}C + T^{1-\beta}K_2^2)^{-1/2}$, the following estimate holds,

$$P\left(\sup_{0 \le t \le T} \left|\int_0^t Z_j(t, s) dW_s^j\right| > L, \quad \|Z\|_\infty \le K_2, \quad \xi \le C\right) \le \exp\left(-\frac{L^2 K_3}{TK_2^2 + T^\alpha C}\right).$$

We now discuss the theorems and concepts needed for their result. To change the measure, Radon-Nikodym theorem is an important tool in the deterministic setting. In the stochastic setting, the Girsanov's transformation theorem, as stated below, is often applied. F. Klebaner offers a good study of this theorem and different techniques regarding change in measure in Chapter 10 of [22].

Theorem 8.7 *(Girsanov's transformation theorem)(see for example Theorem 10.14 in [11]) Let $H_0 = Q^{1/2}H$ and $\psi(\,\cdot\,)$ be an H_0-valued $\mathcal{F}_t$-predictable process such that,*

$$\mathbb{E}\left(e^{\int_0^T \langle \psi(s), dW(s)\rangle_0 - \frac{1}{2}\int_0^T |\psi(s)|_0^2 ds}\right) = 1,$$

for a Q-Weiner process, W, then

$$\widetilde{W}(t) := W(t) - \int_0^t \psi(s) ds,$$

is a Q-Wiener process with respect to $\{\mathcal{F}_t\}_t$ on the probability space, $(\Omega, \mathcal{F}, \widetilde{P})$ with

$$d\widetilde{P}(\omega) = e^{\int_0^T \langle \psi(s), dW(s)\rangle_0 - \frac{1}{2}\int_0^T |\psi(s)|_0^2 ds} dP(\omega).$$

Recall that for a martingale, $\{M_t\}_t$, we have $\mathbb{E}(M_t) = \mathbb{E}(M_0)$. Novikov condition states that for $f(t,\omega) \in \mathbb{R}^n$,

$$Z_t = \exp\left(\int_0^t f(s) dW(s) - \frac{1}{2}\int_0^t f(s)^2 ds\right),$$

is a martingale. Thus, $\mathbb{E}(Z_t) = \mathbb{E}(Z_0) = 1$. Hence, for the result in [25], letting $\psi(t) = \frac{1}{\varepsilon}\dot{g}^j(s)$, we have,

$$\mathbb{E}\left(\exp\left(\frac{1}{\varepsilon}\int_0^T \dot{g}^j(s) dW_s^j - \frac{1}{2\varepsilon^2}\int_0^T |\dot{g}^j(s)|^2 ds\right)\right) = 1,$$

and we obtain by Girsanov's transformation theorem that $\widetilde{W}_t^\varepsilon = W_t - \frac{1}{\varepsilon}\int_0^t \dot{g}^j(s) ds$ is a Wiener process under probability P^ε where,

$$dP^\varepsilon = \exp\left(\frac{1}{\varepsilon}\int_0^T \dot{g}^j(s) dW_s^j - \frac{1}{2\varepsilon^2}\int_0^T |\dot{g}^j(s)|^2 ds\right) dP.$$

Using this fact, the authors introduce the process,

$$Y_t^\varepsilon = x_0 + \int_0^t b(t, s, Y_s^\varepsilon) ds + \int_0^t \varepsilon \sigma_j(t, s, Y_s^\varepsilon) dW_s^j + \int_0^t \sigma_j(t, s, Y_s^\varepsilon)\dot{g}_s^j ds, \quad (8.50)$$

which is the process X_t^ε with the extra term, $\int_0^t \sigma_j(t, s, Y_s^\varepsilon)\dot{g}_s^j ds$. Thus, we let $Y^\varepsilon(\widetilde{W}^\varepsilon) = X^\varepsilon(\widetilde{W}^\varepsilon + \frac{1}{\varepsilon}g)$ to obtain,

$$P^\varepsilon\left(\|X^\varepsilon - S(g)\|_\infty > \rho, \|\varepsilon W - g\|_\infty < \eta\right) = P\left(\|Y^\varepsilon - S(g)\|_\infty > \rho, \|\varepsilon W\|_\infty < \eta\right). \quad (8.51)$$

However, note that we need an exponential bound of,

$$P\left(\|X^\varepsilon - S(g)\|_\infty > \rho, \ \|\varepsilon W - g\|_\infty < \eta\right).$$

To relate this probability with that of P^ε, we let $U^\varepsilon := \exp\left(-\frac{1}{\varepsilon}\int_0^T \dot{g}^j(s)dW_s^j\right)$ and find,

$$P\left(\|X^\varepsilon - S(g)\|_\infty > \rho, \ \|\varepsilon W - g\|_\infty < \eta\right)$$

$$\leq P\left(\|X^\varepsilon - S(g)\|_\infty > \rho, \ \|\varepsilon W - g\|_\infty < \eta, \ U^\varepsilon \leq \exp\left(\frac{\lambda}{\varepsilon^2}\right)\right)$$

$$+ P\left(\|X^\varepsilon - S(g)\|_\infty > \rho, \ \|\varepsilon W - g\|_\infty < \eta, \ U^\varepsilon > \exp\left(\frac{\lambda}{\varepsilon^2}\right)\right)$$

$$\leq \exp\left(\frac{K}{\varepsilon^2}\right) P^\varepsilon\left(\|X^\varepsilon - S(g)\|_\infty > \rho, \ \|\varepsilon W - g\|_\infty < \eta\right) + P\left(U^\varepsilon > \exp\left(\frac{\lambda}{\varepsilon^2}\right)\right).$$

Now by Theorem 8.5,

$$P\left(U^\varepsilon > \exp\left(\frac{\lambda}{\varepsilon^2}\right)\right) = P\left(-\frac{1}{\varepsilon}\int_0^T \dot{g}_s^j dW_s^j > \frac{\lambda}{\varepsilon^2}\right) = P\left(\left|\int_0^T \dot{g}_s^j dW_s^j\right| > \frac{\lambda}{\varepsilon}\right)$$

$$\leq \exp\left(-\frac{K}{\varepsilon^2}\right).$$

Therefore,

$$P\left(\|X^\varepsilon - S(g)\|_\infty > \rho, \ \|\varepsilon W - g\|_\infty < \eta\right)$$

$$\leq \exp\left(\frac{K}{\varepsilon^2}\right) P^\varepsilon\left(\|X^\varepsilon - S(g)\|_\infty > \rho, \ \|\varepsilon W - g\|_\infty < \eta\right) + \exp\left(-\frac{K}{\varepsilon^2}\right), \quad (8.52)$$

and because of (8.51) it is sufficient to show,

$$P\left(\|Y^\varepsilon - S(g)\|_\infty > \rho, \ \|\varepsilon W\|_\infty < \eta\right) \leq \exp\left(-\frac{R}{\varepsilon^2}\right), \quad (8.53)$$

and afterwards group the two exponential bounds on the right hand side of (8.52).

Theorem 8.8 (*Theorem 1 in [25]*) *Family* $\{X^\varepsilon\}_{0<\varepsilon\leq 1}$ *satisfies the large deviations principle in* $\mathcal{C}(0, T; \mathbb{R}^d)$ *with rate function,*

$$I(\phi) = \inf\left\{\frac{1}{2}\int_0^T |\dot{g}_s|^2 ds : \ \phi = S(g), \ g \in \mathcal{H}\right\}. \quad (8.54)$$

Proof. Recall that to prove the large deviation principle by the Azencott method, we need,

  i. another process that is known to satisfy the large deviation principle,

 ii. show that the Friedlin-Wentzell inequality holds,

iii. prove the continuity of the map, $h \to S(h)$.

For the first condition, the authors let $\{\varepsilon W(\,\cdot\,)\}_{\varepsilon > 0}$ be the other process, which is known by the Schilder's theorem to satisfy the large deviation principle with rate function, $I(\phi)$ in (8.54) in the uniform topology (see for example Section 5.2 in [14]). In addition, condition (*iii*) is shown in the article's Lemma 3 by proving that for any fixed $\varepsilon > 0$ and $h, g \in \mathcal{H}$,

$$\|S(g) - S(h)\|_\infty \leq 2a\sqrt{\varepsilon} + K\|h - g\|_\infty.$$

Thus, the main focus of the paper lies in achieving the Freidlin-Wentzell inequality (8.53). Following the usual approach in the literature, to achieve (8.53), time discretization of the process $\{Y^\varepsilon\}_\varepsilon$ is applied. More precisely, we divide the time interval $[0, T]$ into disjoint intervals, $\Delta_\ell := [t_i^n, t_{i+1}^n)$, where $n \in \mathbb{N}, i = 0, 1, ..., n$ and $t_i^n = \left(\frac{T}{n}\right)i$. The time discretized version of Y^ε is then denoted as $Y^{\varepsilon,n}$. Using the process, $\{Y^{\varepsilon,n}\}_{0 < \varepsilon \leq 1}$ note that inequality (8.53) is implied by establishing the following two inequalities,

$$P\left(\|Y^\varepsilon - Y^{\varepsilon,n}\|_\infty > \mu\right) \leq \exp\left(-\frac{R}{\varepsilon^2}\right), \tag{8.55}$$

$$P\left(\|Y^\varepsilon - S(g)\|_\infty > \rho, \ \|\varepsilon W\|_\infty < \eta, \ \|Y^\varepsilon - Y^{\varepsilon,n}\|_\infty \leq \mu\right) \leq \exp\left(-\frac{R}{\varepsilon^2}\right). \tag{8.56}$$

For example for (8.55), for fixed n, we let $t \in \Delta_i^n$ and find,

$$
\begin{aligned}
Y^\varepsilon - Y^{\varepsilon,n} &= \int_0^t b(t, s, Y_s^\varepsilon)ds - \int_0^{t_i^n} b(t_i^n, s, Y_s^\varepsilon)ds \\
&+ \int_0^t \varepsilon\sigma_j(t, s, Y_s^\varepsilon)dW_s^j - \int_0^t \varepsilon\sigma_j(t_i^n, s, Y_s^\varepsilon)dW_s^j \\
&+ \int_0^t \sigma_j(t, s, Y_s^\varepsilon)\dot{g}_s^j ds - \int_0^{t_i^n} \sigma_j(t_i^n, s, Y_s^\varepsilon)\dot{g}_s^j ds.
\end{aligned}
$$

Notice that since the coefficients b and σ are bounded and α-Hölder continuous with respect to time, we have, by adding and subtracting $\int_0^{t_i^n} b(t, s, Y_s^\varepsilon)ds$,

$$\int_0^t b(t,s,Y_s^\varepsilon)ds - \int_0^{t_i^n} b(t_i^n,s,Y_s^\varepsilon)ds = \int_{t_i^n}^t b(t,s,Y_s^\varepsilon)ds$$

$$+ \int_0^{t_i^n} \left(b(t,s,Y_s^\varepsilon) - b(t_i^n,s,Y_s^\varepsilon) \right) ds$$

$$\leq K|t - t_i^n| + K|t - t_i^n|^\alpha \, t_i^n,$$

where $0 < \alpha \leq 1$. Similarly, with the help of the Cauchy-Schwarz inequality,

$$\int_{t_i^n}^t \sigma_j(t,s,Y_s^\varepsilon) \, \dot{g}_s^j ds + \int_0^{t_i^n} \left(\sigma_j(t,s,Y_s^\varepsilon) - \sigma_j(t_i^n,s,Y_s^\varepsilon) \right) \dot{g}_s^j ds$$

$$\leq K\sqrt{M}|t - t_i^n|^2 + Kt_i^n \sqrt{M}|t - t_i^n|^\alpha.$$

Hence, we have,

$$\left| Y_t^\varepsilon - Y_t^{\varepsilon,n} \right| \leq K|t - t_i^n| + K|t - t_i^n|^\alpha \, t_i^n + K\sqrt{M}|t - t_i^n|^2 + Kt_i^n \sqrt{M}|t - t_i^n|^\alpha$$

$$+ \left| \varepsilon \int_{t_i^n}^t \sigma_j(t,s,Y_s^\varepsilon)dW_s^j \right| + \left| \varepsilon \int_0^{t_i^n} \left(\sigma_j(t,s,Y_s^\varepsilon) - \sigma_j(t_i^n,s,Y_s^\varepsilon) \right) dW_s^j \right|.$$

Recall that for random variables, $X_1, X_2, ..., X_n$ and a constant $k > 0$,

$$P\left((X_1 + X_2 + ... + X_n) \geq k \right) \leq P\left(X_1 \geq \frac{k}{n} \right) + P\left(X_2 \geq \frac{k}{n} \right) + ... + P\left(X_n \geq \frac{k}{n} \right).$$

Then using the infinity norm, we have,

$$P\left(\left\| Y^\varepsilon - Y^{\varepsilon,n} \right\|_\infty > \mu \right) = \sum_{i=1}^n P\left(\sup_{t \in \Delta_i^n} \left\| Y^\varepsilon - Y^{\varepsilon,n} \right\|_\infty > \mu \right)$$

$$\leq \sum_{i=1}^n P\left(\sup_{t \in \Delta_i^n} K|t - t_i^n| + K|t - t_i^n|^\alpha \, t_i^n + K\sqrt{M}|t - t_i^n|^2 + Kt_i^n \sqrt{M}|t - t_i^n|^\alpha > \frac{\mu}{3} \right)$$

$$+ \sum_{i=1}^n P\left(\sup_{t \in \Delta_i^n} \left| \varepsilon \int_{t_i^n}^t \sigma_j(t,s,Y_s^\varepsilon)dW_s^j \right| > \frac{\mu}{3} \right)$$

$$+ \sum_{i=1}^n P\left(\sup_{t \in \Delta_i^n} \left| \varepsilon \int_0^{t_i^n} \left(\sigma_j(t,s,Y_s^\varepsilon) - \sigma_j(t_i^n,s,Y_s^\varepsilon) \right) dW_s^j \right| > \frac{\mu}{3} \right).$$

Theorem 8.6 is then applied to the last two probabilities above to obtain the required exponential bound, since their integrands involving $\varepsilon \sigma_j(t,s,Y_s^\varepsilon)$ are bounded. As for the first probability, since n was arbitrary, it can be chosen to be sufficiently large to ensure that $|t - t_i^n|$ in each term on the left hand side of the inequality is small enough and this probability become zero for any $\mu > 0$. Similarly, by applying Theorem 8.6, inequality (8.56) is achieved. □

As for the weak convergence approach, manuscripts [5,16] offer in depth studies of the approach and describe large deviations results for various stochastic processes such as random walk and Markov processes. The procedure on how to apply this method to prove large deviations for SPDEs was introduced by A. Budhiraja, P. Dupuis and V. Maroulas in [6] in 2008, which was based on the variational representations provided in [4]. More precisely, A. Budhiraja and P. Dupuis proved for a bounded, Borel measurable function, $f : \mathcal{C}([0, T]; \mathbb{R}^\infty) \to \mathbb{R}$,

$$-\log \mathbb{E}\left(\exp\left(-f(\beta)\right)\right) = \inf_{u \in \mathcal{P}(\ell_2)} \mathbb{E}\left(\frac{1}{2}\int_0^T \|u(s)\|_{\ell_2}^2 ds + f\left(\beta + \int_0^{\cdot} u(s)ds\right)\right),$$

$$(8.57)$$

where $\beta = \{\beta_j\}_{j \geq 1}$ is an infinite sequence of independent, real, standard Brownian motions and $\ell_2 := \{x = (x_1, x_2, \ldots) : x_i \in \mathbb{R} \text{ and } \sum_i x_i^2 < \infty\}$. Note that the above is the form one needs for the left hand side of the bounds in Laplace principle and it may be used to achieve these bounds in the case of Polish spaces to obtain large deviations. This type of representation was also shown for a Q-Wiener process, W, and Brownian sheet, B in [4] and they were then used in [6] to obtain large deviations for processes depending on each of these types of noise. Namely, large deviations for

$$X_1^{\varepsilon,x} := \mathcal{G}^\varepsilon\left(x, \sqrt{\varepsilon}W\right), \quad X_2^{\varepsilon,x} := \mathcal{G}\left(x, \sqrt{\varepsilon}\beta\right), \quad \text{and} \quad X_3^{\varepsilon,x} := \mathcal{G}^\varepsilon\left(x, \sqrt{\varepsilon}B\right),$$

were established in [6] in its Theorems 5, 6 and 7, respectively. Due to the similarity of the conditions required in each of the three theorems, we focus on Theorem 5. Suppose W is a Q-Wiener process taking values in space H and denote $H_0 = Q^{1/2}H$. Let $\mathcal{E}_0$ and $\mathcal{E}$ be two Polish spaces and $\mathcal{G}^\varepsilon : \mathcal{E}_0 \times \mathcal{C}([0, T]; H) \to \mathcal{E}$ be a map, $\mathcal{G}^\varepsilon(x, \sqrt{\varepsilon}W)$. Similar to (8.46), they use the space given by,

$$S_M(H_0) = \left\{h \in L^2([0, T]; H_0) \text{ predictable} : \int_0^T \|h(s)\|_0^2 ds \leq M\right\}, \quad (8.58)$$

where $L^2(0, T; H_0)$ is composed of functions, $f : [0, T] \to H_0$ in L^2. A process $\{X_t\}_{t \geq 0}$ is called predictable with respect to the filtration $\mathcal{F}$ if for every t, X_t is $\mathcal{F}_{t-1}$- measurable. Theorem 5 gives the following two conditions that ensure that the process $\{X_1^{\varepsilon,x}\}_{\varepsilon>0}$ defined above, satisfies the large deviation principle.

i. There exists a measurable map, $\mathcal{G}^0 : \mathcal{E}_0 \times \mathcal{C}([0,T]; H) \to \mathcal{E}$ such that for each constant $M < \infty$ and compact set K in $\mathcal{E}_0$, the set,

$$\Gamma_{M,K} = \left\{ \mathcal{G}^0 \left(x, \int_0^\cdot h(s)ds \right) : h \in S_M(H_0), \ x \in K \right\}, \tag{8.59}$$

is compact in $\mathcal{E}$,

ii. If $\{h^\varepsilon\}_{\varepsilon>0}$ is a family in $S_M(H_0)$ for $M < \infty$ such that $h^\varepsilon \xrightarrow{d} h$ as $\varepsilon \to 0$ and $\{x^\varepsilon\}_{\varepsilon>0}$ is a family in $\mathcal{E}_0$ such that $x^\varepsilon \to x$ as $\varepsilon \to 0$, then,

$$\mathcal{G}^\varepsilon \left(x^\varepsilon, \sqrt{\varepsilon}W(\cdot) + \int_0^\cdot h^\varepsilon(s)ds \right) \xrightarrow{d} \mathcal{G}^0 \left(x, \int_0^\cdot h(s)ds \right), \tag{8.60}$$

as $\varepsilon \to 0$, where $\xrightarrow{d}$ denotes the convergence in distribution.

If the above two conditions hold then the large deviations principle is achieved for $\{X_1^{\varepsilon,x}\}_{\varepsilon>0}$ with rate function,

$$I(u) = \inf_{\{h\in L^2([0,T];H_0) : \ u = \mathcal{G}^0(x, \ \int_0^\cdot h(s)ds)\}} \frac{1}{2} \int_0^T \|h(s)\|_0^2 ds, \tag{8.61}$$

for $x \in \mathcal{E}_0$ and $f \in \mathcal{E}$.

In the context of SPDEs:

$\mathcal{G}^\varepsilon(x, \sqrt{\varepsilon}W)$ is the unique solution to the original SPDE,

$\mathcal{G}^0 \left(x, \int_0^\cdot h(s)ds \right)$ is the unique solution to the controlled PDE obtained by replacing the noise in the original equation by $h(t) \in S_M(H_0)$,

$\mathcal{G}^\varepsilon \left(x^\varepsilon, \sqrt{\varepsilon}W(\cdot) + \int_0^\cdot h^\varepsilon(s)ds \right)$, referred to as the stochastic controlled equation, is the unique solution to the equation formed by adding the controlled term to the original SPDE.

Thus, one is required to prove that the solutions to the controlled PDE form a compact set and that the stochastic controlled equation converges in distribution to the controlled PDE.

To apply the above theorem, the first step is to ensure the well-posedness of the original SPDE, controlled PDE and the stochastic controlled PDE. For the three equations, the well-posedness of solutions in the form of mild or weak solutions in PDE but strong in probability are sufficient for the result to hold.

In [15], J. Duan and A. Millet establish large deviations for two-dimensional stochastic Boussinesq equation, given by,

$$du^\varepsilon + \left(\nu A_1 u^\varepsilon + B_1(u^\varepsilon, u^\varepsilon) - \theta^\varepsilon e_2\right) dt = \sqrt{\varepsilon}\sigma_1(t, \phi^\varepsilon) dW^1(t), \qquad (8.62)$$
$$d\theta^\varepsilon + \left(\kappa A_2 \theta^\varepsilon + B_2(u^\varepsilon, \theta^\varepsilon) - u_2^\varepsilon\right) dt = \sqrt{\varepsilon}\sigma_2(t, \phi^\varepsilon) dW^2(t),$$

where W_1 and W_2 are Q-Wiener processes and θ^ε represents the temperature of the fluid at time t and position x. As may be observed from above, Boussinesq equation is Navier-Stokes equation coupled with the equation representing the change in temperature. In [15], the authors group the above two equations to give the following abstract form,

$$d\phi^\varepsilon + \left(A\phi^\varepsilon + B(\phi^\varepsilon) + R(\phi^\varepsilon)\right) dt = \sqrt{\varepsilon}\sigma(t, \phi^\varepsilon) dW(t), \qquad (8.63)$$
$$\phi^\varepsilon(0) = (u_0^\varepsilon, \theta_0^\varepsilon),$$

where $\phi^\varepsilon = (u^\varepsilon, \theta^\varepsilon)$, $A\phi = (\nu A_1 u, \kappa A_2 \theta)$, $B(\phi) = (B_1(u, u), B_2(u, \theta))$, $R\phi = (-\theta e_2, -u_2)$, and $\sigma(t, \phi) = (\sigma_1(t, \phi), \sigma_2(t, \phi))$. The controlled PDE is given by,

$$d\phi(t) + (A\phi(t) + B(\phi(t)) + R\phi(t)) dt = \tilde{\sigma}(\phi(t))h(t)dt,$$
$$\phi(0) = \xi, \qquad (8.64)$$

where $h \in S_M(H_0)$. Also the stochastic controlled equation is as follows,

$$d\phi_h^\varepsilon(t) + \left(A\phi_h^\varepsilon(t) + B(\phi_h^\varepsilon(t)) + R\phi_h^\varepsilon(t)\right) dt = \sqrt{\varepsilon}\sigma(\phi_h^\varepsilon(t))dW(t) + \tilde{\sigma}(\phi_h^\varepsilon(t))h(t)dt, \qquad (8.65)$$

for which their result of well-posedness of weak solutions in PDE sense and strong in probability were discussed in the previous section. The authors prove the existence and uniqueness of solutions to (8.65) and from it derive the well-posedness of (8.63) and (8.64). Afterwards, the two conditions of weak convergence approach are proved in the paper's Propositions 4.3 and 4.4. For the two conditions, their proof is similar to that of [9], which is described in detail in the next section.

8.3 Large Deviations for Stochastic Navier Stokes Equation

Here we present some results on large deviations for two-dimensional stochastic incompressible Navier-Stokes equation that apply the weak convergence approach. As in previous sections, we follow the convention of $|\cdot|$ and $\|\cdot\|$ denoting norms in H and V, respectively.

In [9], I. Chueshov and A. Millet form the following class of SPDEs,

$$du^\varepsilon(t) + \left(Au^\varepsilon(t) + B(u^\varepsilon(t)) + \widetilde{R}(t, u^\varepsilon(t))\right) dt = \sqrt{\varepsilon}\sigma(t, u^\varepsilon(t)) dW(t), \quad (8.66)$$

from which the four fluid models: Navier-Stokes, Boussinesq, MHD and Shell model equations may be obtained as special cases. In (8.66), operators A and B are the same as those defined in Section 8.1. W is a Q-Wiener process and for a Hilbert space, H, map $\widetilde{R} : [0, T] \times H \to H$ is continuous satisfying, for all $u, v \in H$ and $t \in [0, T]$,

$$|\widetilde{R}(t, 0)| \le K, \quad \text{and} \quad |\widetilde{R}(t, u) - \widetilde{R}(t, v)| \le K|u - v|. \quad (8.67)$$

In addition, assumptions on $\sigma(t, u)$ are given as follows,

$$\|\sigma(t, u)\|_{L_Q}^2 \le K_1 + K_2|u|^2 + K_3\|u\|^2 \quad \text{and}$$

$$\|\sigma(t, u) - \sigma(t, v)\|_{L_Q}^2 \le K_1|u - v|^2 + K_2\|u - v\|^2, \quad (8.68)$$

for all $u, v \in V$ and $t \in [0, T]$. With $H_0 = Q^{1/2}H$, the space of controls is,

$$S_M = \left\{ h \in L^2(0, T; H_0) \text{ predictable} : \int_0^T |h(s)|_0^2 ds \le M \right\}, \quad (8.69)$$

with which the controlled PDE and the stochastic controlled equations are formed by,

$$du_h(t) + \left(Au_h(t) + B(u_h(t)) + \widetilde{R}(t, u_h(t))\right) dt = \sigma(t, u_h(t))h(t) dt, \quad (8.70)$$

and

$$du_h^\varepsilon(t) + \left(Au_h^\varepsilon(t) + B(u_h^\varepsilon(t)) + \widetilde{R}(t, u_h^\varepsilon(t))\right) dt$$
$$= \sqrt{\varepsilon}\sigma(t, u_h^\varepsilon(t)) dW(t) + \widetilde{\sigma}(t, u_h^\varepsilon(t))h(t) dt, \quad (8.71)$$

respectively. To achieve the well-posedness of solutions to equations (8.66), (8.70) and (8.71) in the space of study, $X = \mathcal{C}([0, T]; H) \cap L^2(0, T; V)$, the well-posedness of (8.71) is proved in the paper's Theorem 2.4, from which the well-posedness of the other two equations were deduced. Following a similar reasoning as in [15] discussed in Section 8.1, the authors show the existence and uniqueness of solutions that are strong in probability and weak in PDE sense and derive the following estimate,

$$\mathbb{E}\left(\sup_{0\leq t\leq T} |u_h(t)|^4 + \int_0^T \|u_h(t)\|^2\, dt + \int_0^T \|u_h(t)\|_{\mathcal{H}}^4\, dt\right) \leq K\left(1 + \mathbb{E}|\xi|^4\right),$$

$$(8.72)$$

where $\xi := u_h(0) \in H$ and $\mathcal{H} := \mathrm{Dom}(A^s)$ with $s \in \left[0, \frac{1}{4}\right]$. For their large deviation result, they prove the following lemma and use the Skorohod representation theorem, also referred to as the Skorohod embedding theorem, stated below.

Lemma 8.9 *(Lemma 3.3 in [9]) Let,*

$$G_N(t) := \left\{\omega : \max\left\{\sup_{0\leq s\leq t} |u_h^\varepsilon(s)(\omega)|^2, \int_0^t \|u_h^\varepsilon(s)(\omega)\|^2\, ds\right\} \leq N\right\}, \quad (8.73)$$

then for a map $\psi_n : [0, T] \to [0, T]$ satisfying, for every $s \in [0, T]$ and some constant $c > 0$,

$$s \leq \psi_n(s) \leq \min\left\{(s + c2^{-n}), T\right\}, \quad (8.74)$$

the following estimate holds,

$$\mathbb{E}\left(1_{G_N(T)} \int_0^T |u_h^\varepsilon(s) - u_h^\varepsilon(\psi_n(s))|^2\, ds\right) \leq K\, 2^{-\frac{n}{2}}. \quad (8.75)$$

Theorem 8.10 *(Skorohod Representation Theorem)* In a probability space, $(\Omega, \mathcal{F}, P)$, for a sequence of random variables, $\{X_n\}_n$ taking values in a Polish space, $\mathcal{X}$, if $X_n \xrightarrow{d} X$, then there exists another probability space, $(\tilde{\Omega}, \tilde{\mathcal{F}}, \tilde{P})$ in which there are random variables, $\{\tilde{X}_n\}_n$ and $\tilde{X}$ taking values in $\mathcal{X}$ such that, for each n, $X_n \overset{d}{=} \tilde{X}_n$, $X \overset{d}{=} \tilde{X}$, and $\tilde{X}_n \to \tilde{X}$ $\tilde{P}$- a.s.

For a proof of the Skorohod representation theorem see for example Theorem 2.4 in [11]. In addition, [16] offers many examples of the application of this theorem to different processes such as Markov chain and random walk. For the proof of the main result, discussed below, we also recall that a linear map, $T : X \to Y$ is a compact operator if for every open set $U \subset X$, the set $\overline{T(U)}$ is compact in Y. Here $\overline{B}$ denotes the closure of set B. It is known that if g is a Hilbert-Schmidt operator, then it is also a compact operator. By definition, compact operator T is bounded and if $h \in X$, then $\|T_n h - Th\| \to 0$ as $n \to \infty$. Also if $\{x_n\}_n$ is a bounded sequence in X, then it has a subsequence, $\{\tilde{x}_k\}_k$ such that for each fixed n, $\{T_n \tilde{x}_k\}_k$ converges in Y as $k \to \infty$. For proofs of these facts on compact operators, we refer the reader to Section 1.2 of [20].

Theorem 8.11 (*Theorem 3.2 in [9]*) *The family*, $\{u^\varepsilon\}_{\varepsilon>0}$ *of solutions of* (8.66) *satisfies the large deviations principle in* $\mathcal{C}(0,T;H) \cap L^2(0,T;V)$ *with rate function*,

$$I_\xi(u) = \lim_{\{h \in L^2(0,T;H_0) \,:\, u = \mathcal{G}^0(\int_0^\cdot h(s)ds)\}} \frac{1}{2} \int_0^T |h(s)|_0^2 ds. \tag{8.76}$$

Proof. Recall that to obtain the large deviation principle by the weak convergence approach, one needs to verify the following two conditions,

i. the set, $K_M = \{\mathcal{G}^0(x, \int_0^\cdot h(s)ds) : h \in S_M\}$ is compact, where $\mathcal{G}^0$ represents the unique solution to the controlled PDE,

ii. the stochastic controlled equation converges in distribution to the controlled PDE as $\varepsilon \to 0$.

In [9] the above two conditions are proved in Propositions 3.5 and 3.4, respectively.

For the proof of condition (i), recall that in a metric space, a set is compact if and only if it is sequentially compact. Hence, the authors prove that K_M is sequentially compact as follows. Let $\{h_n\}_n$ be a sequence in S_M. As noted in Section 4 of [6], S_M is a compact Polish space with metric,

$$d(x,y) = \sum_{k=1}^\infty \frac{1}{2^k} \left| \int_0^T \langle x(s) - y(s), e_k(s) \rangle_0 \, ds \right|,$$

for an orthonormal basis, $\{e_k\}_{k\geq 1}$ of H_0. Therefore, S_M is closed and bounded in $L^2(0,T;H_0)$, which implies that it is relatively compact. That is, there is a subsequence, $\{h_{n_k}\}_{k\geq 1}$ such that $h_{n_k} \xrightarrow{d} \tilde{h} \in S_M$. We let h_{n_k} still be denoted as h_n.

Let $(u_{h_n})_n$ be the sequence of solutions corresponding to the set of controls, $\{h_n\}_n$ found above. We show that $(u_{h_n})_n$ converges strongly in space $\mathcal{C}(0,T;H) \cap L^2(0,T;V)$ to $u_{\tilde{h}}$, which is the solution of the controlled PDE corresponding to $\tilde{h}$, the limit of $\{h_n\}_n$.

Denoting the difference by $U_h^n := u_{h_n} - u_{\tilde{h}}$, similar to (8.35) we apply the Itô's formula in (8.7) to obtain,

$$|U_n(t)|^2 + 2 \int_0^t \|U_n(s)\|^2 \, ds = -2 \int_0^t \Big(B(u_{h_n}(s)) - B(u_{\widetilde{h}}(s)), U_n(s) \Big) ds$$

$$- 2 \int_0^t \Big(\widetilde{R}(s, u_{h_n}(s)) - \widetilde{R}(s, u_{\widetilde{h}}(s)), U_n(s) \Big) ds$$

$$+ 2 \int_0^t \Big(\sigma(s, u_{h_n}(s)) h_n(s) - \sigma(s, u_{\widetilde{h}}(s)) \widetilde{h}(s), U_n(s) \Big) ds$$

$$\leq |I_1| + |I_2| + |I_3|. \tag{8.77}$$

Note that here, there is no noise term and the last two terms in (8.7) may be disregarded. By (8.29),

$$|I_1| \leq \int_0^t \|U_n(s)\|^2 ds + K_1 \int_0^t |U_n(s)|^2 \|u_{\widetilde{h}}(s)\|_{L^4}^4 ds.$$

Using conditions in (8.67) we have,

$$|I_2| \leq 2 \int_0^t |\widetilde{R}(s, u_{h_n}(s)) - \widetilde{R}(s, u_{\widetilde{h}}(s))| \, |U_n(s)| ds \leq 2K_2 \int_0^t |U_n(s)|^2 ds.$$

For the third term, by adding and subtracting $\sigma(s, u_{\widetilde{h}}(s)) h_n(s)$ we write,

$$|I_3| \leq 2 \int_0^t \Big(\big(\sigma(s, u_{h_n}(s)) - \sigma(s, u_{\widetilde{h}}(s)) \big) \, h_n(s), U_n(s) \Big) ds$$

$$+ 2 \int_0^t \Big(\big(h_n(s) - \widetilde{h}(s) \big) \sigma(s, u_{\widetilde{h}}(s)), U_n(s) \Big) ds = I_{31} + I_{32},$$

where the first term may be bounded applying the conditions in (8.68) and Cauchy-Schwarz inequality as follows,

$$I_{31} \leq 2 \int_0^t \|\sigma(s, u_{h_n}(s)) - \sigma(s, u_{\widetilde{h}}(s))\|_{L_Q} \, |h_n(s)|_0 \, |U_n(s)| ds$$

$$\leq 2\sqrt{K_1} \int_0^t |U_n(s)|^2 \, |h_n(s)|_0 \, ds + 2\sqrt{K_2} \int_0^t \|U_n(s)\| \, |h_n(s)|_0 \, |U_n(s)| ds$$

$$\leq 2\sqrt{K_1} \int_0^t |U_n(s)|^2 \, |h_n(s)|_0 \, ds + 2\sqrt{K_2 M} \left(\int_0^t \|U_n(s)\|^2 |U_n(s)|^2 ds \right)^{\frac{1}{2}},$$

where we have noted that $h_n \in S_M$. Taking the supremum of $|U_n(s)|$ outside of the second integral on the right, we obtain by (8.72),

$$I_{31} \leq 2K \int_0^t |U_n(s)|^2 |h_n(s)|_0 \, ds,$$

which may further be simplified as,

$$I_{31} \leq 2K \left(\int_0^t |U_n(s)|^4 ds \right)^{\frac{1}{2}} \left(\int_0^t |h_n(s)|_0^2 ds \right)^{\frac{1}{2}} \leq 2K\sqrt{M} \left(\int_0^t |U_n(s)|^4 ds \right)^{\frac{1}{2}}.$$

Thus, we arrive at,

$$|U_n(t)|^2 + 2 \int_0^t \|U_n(s)\|^2 \, ds$$

$$\leq \int_0^t \|U_n(s)\|^2 ds + \int_0^t |U_n(s)|^2 \left(K_1 \|u_{\widetilde{h}}(s)\|_{L^4}^4 + 2K_2 \right) ds$$

$$+ 2K\sqrt{M} \left(\int_0^t |U_n(s)|^4 ds \right)^{\frac{1}{2}} + 2 \int_0^t \left(\sigma(s, u_{\widetilde{h}}(s)) \left(h_n(s) - \widetilde{h}(s) \right), U_n(s) \right) ds.$$

Grouping the term $\int_0^t \|U_n(s)\|^2 ds$ from both sides and applying estimate (8.72), we attain,

$$|U_n(t)|^2 + \int_0^t \|U_n(s)\|^2 ds \leq K + 2 \int_0^t \left(\sigma(s, u_{\widetilde{h}}(s)) \left(h_n(s) - \widetilde{h}(s) \right), U_n(s) \right) ds.$$

Now for $N \in \mathbb{N}$ and $s \in [kT2^{-N}, (k+1)T2^{-N}) \subset [0, T]$, let $\bar{s}_N := (k+1)T2^{-N}$ be a step function mapping $s \mapsto \bar{s}_N$. Note that by adding and subtracting the term,

$$\int_0^T \left((\sigma(s, u_{\widetilde{h}}(s)) + \sigma(\bar{s}_N, u_{\widetilde{h}}(s)) + \sigma(\bar{s}_N, u(\bar{s}_N))) (h_n(s) - \widetilde{h}(s)), U_n(\bar{s}_N) \right) ds,$$

we have for $t_k = kT2^{-N}$ with $0 \leq k \leq 2^N$,

$$\int_0^T \left| \left(\sigma(s, u_{\widetilde{h}}(s)) \left(h_n(s) - \widetilde{h}(s) \right), U_n(s) \right) \right| ds$$

$$= \int_0^T \left| \left(\sigma(s, u_{\widetilde{h}}(s)) \left(h_n(s) - \widetilde{h}(s) \right), U_n(s) - U_n(\bar{s}_N) \right) \right| ds$$

$$+ \int_0^T \left| \Big(\big(\sigma(s, u_{\widehat{h}}(s)) - \sigma(\overline{s}_N, u_{\widehat{h}}(s)) \big) (h_n(s) - \widetilde{h}(s)), U_n(\overline{s}_N) \Big) \right| ds$$

$$+ \int_0^T \left| \Big(\big(\sigma(\overline{s}_N, u_{\widehat{h}}(s)) - \sigma(\overline{s}_N, u(\overline{s}_N)) \big) (h_n(s) - \widetilde{h}(s)), U_n(\overline{s}_N) \Big) \right| ds$$

$$+ \sup_{1 \le k \le 2^N} \lim_{t_{k-1} \le t \le t_k} \left| \Big(\sigma(t_k, u_{\widehat{h}}(t_k)) \int_{t_{k-1}}^t \big(h_n(s) - \widetilde{h}(s) \big) ds, U_n(t_k) \Big) \right|$$

$$+ \sum_{k=1}^{2^N} \Big(\sigma(t_k, u_{\widehat{h}}(t_k)) \int_{t_{k-1}}^{t_k} \big(h_n(s) - \widetilde{h}(s) \big) ds, U_n(t_k) \Big)$$

$$= J_1 + J_2 + J_3 + J_4 + J_5. \tag{8.78}$$

Notice that for the last term, J_5, since $\sigma(t_k, u_{\widehat{h}}(t_k)) \in L_Q(H_0; H)$, then it is a Hilbert-Schmidt and hence a compact operator. Since $\{h_n\}_n$ is a subsequence that converges to h in S_M, we have by the dominated convergence theorem, $\int_{t_{k-1}}^{t_k} \big(h_n(s) - \widetilde{h}(s) \big) ds \to 0$ and thus we obtain, for each k in the sum in J_5, the strong convergence of $\sigma(t_k, u_{\widehat{h}}(t_k)) \int_{t_{k-1}}^{t_k} \big(h_n(s) - \widetilde{h}(s) \big) ds$ to zero, which then by the Cauchy-Schwarz inequality and estimate (8.72) gives $J_5 \to 0$. To apply this reasoning to J_4, we fix the bounds on its integral as follows for the dominated convergence theorem to hold,

$$J_4 \le \sup_{1 \le k \le 2^N} \left| \Big(\sigma(t_k, u_{\widehat{h}}(t_k)) \int_{t_{k-1}}^{t_k} \big(h_n(s) - \widetilde{h}(s) \big) ds, U_n(t_k) \Big) \right|. \tag{8.79}$$

Observe that in (8.79), $\sigma(t_k, u_{\widehat{h}}(t_k))$ depends on k, which is not fixed and hence the above technique for J_5 cannot be applied. Instead, with the help of the Cauchy-Schwarz inequality and the condition in (8.68) we have,

$$J_4 \le \sup_{1 \le k \le 2^N} \big| \sigma(t_k, u_{\widehat{h}}(t_k)) \big|_{L_Q} \int_{t_{k-1}}^{t_k} \big| h_n(s) - \widetilde{h}(s) \big|_0 ds \, |U_n(t_k)|$$

$$\le K \sup_{1 \le k \le 2^N} \int_{t_{k-1}}^{t_k} \big| h_n(s) - \widetilde{h}(s) \big|_0 ds \le K 2^{-\frac{N}{4}},$$

where the convergence of h_n to $\widetilde{h}$ and the dominated convergence theorem were used. As for the other three terms, we let $\psi_N(s) := \overline{s}_N$ and note that,

$$s \le (k+1) T^{-N} = \overline{s}_N \le s + T 2^{-N},$$

and also $(k+1) T 2^{-N} \le s + T 2^{-N}$ based on the bound on s. Thus, with $c = T$ in (8.74), the condition on $\psi_N(s)$ is satisfied and Lemma 8.1 may be applied along with the condition in (8.68), Cauchy-Schwarz inequality and estimate (8.72) to bound the first three terms by $K2^{-\frac{N}{4}}$. Therefore, we achieve,

$$|U_n(t)|^2 + \int_0^t \|U_n(s)\|^2 ds \leq \int_0^t |U_n(s)|^2 \left(K_1 \|u_{\tilde{h}}(s)\|_{L^4}^4 + 2K_2\right) ds + K2^{-\frac{N}{4}},$$

which by Grownwall's inequality becomes,

$$|U_n(t)|^2 + \int_0^t \|U_n(s)\|^2 ds \leq K2^{-\frac{N}{4}} \exp\left(K_1 \|u_{\tilde{h}}(s)\|_{L^4}^4 + 2K_2\right).$$

Setting $N \to \infty$ and noting that by (8.72), $\|U_n(s)\|_{L^4}^4 < \infty$, we obtain the strong convergence of $u_n(t)$ to $u_{\tilde{h}}(t)$ in $\mathcal{C}([0,T]; H) \cap L^2(0,T; V)$. Since for any sequence $\{u_n\}_n$ in K_M, we may find a subsequence $\{u_{n_k}\}_k$ that converges to a u_h in K_M by the above reasoning, then we have that K_M is sequentially compact.

As for the second condition for large deviations, the authors prove that $u_{h_\varepsilon} - u_h$ converges to zero in probability, which implies its convergence in distribution. Therefore, setting, $U^\varepsilon = u_{h_\varepsilon} - u_h$, the goal is to prove that for every $\delta > 0$,

$$\lim_{\varepsilon \to 0} P\left(\|U^\varepsilon\|_{\mathcal{C}([0,T];H) \cap L^2(0,T;V)} > \delta\right) = 0. \tag{8.80}$$

Since S_M is a Polish space and $h_\varepsilon \overset{d}{\to} h$ in S_M, then by Skorohod representation theorem, there exists $(\widetilde{\Omega}, \widetilde{\mathcal{F}}, \widetilde{P})$ and $\{\widetilde{h}_\varepsilon\}$ and $\widetilde{h}$ in this space such that $\widetilde{h}_\varepsilon \overset{d}{=} h_\varepsilon$, $\widetilde{h} \overset{d}{=} h$ and $\widetilde{h}_\varepsilon \to \widetilde{h}$ $\widetilde{P}$-a.s. as $\varepsilon \to 0$. Thus, we obtain the stronger a.s. convergence in $(\widetilde{\Omega}, \widetilde{\mathcal{F}}, \widetilde{P})$ and since $\widetilde{h}_\varepsilon, \widetilde{h} \in S_M$, we have by the dominated convergence theorem, $\int_0^t \widetilde{h}_\varepsilon(s)ds \to \int_0^t \widetilde{h}(s)ds$ $\widetilde{P}$-a.s. for fixed $t \in [0,T]$ as $\varepsilon \to 0$. Since $\widetilde{h}_\varepsilon \overset{d}{=} h_\varepsilon$ and $\widetilde{h} \overset{d}{=} h$, then $\mathbb{E}_{\widetilde{p}}(\widetilde{h}_\varepsilon) = \mathbb{E}_P(h_\varepsilon)$ and $\mathbb{E}_{\widetilde{p}}(\widetilde{h}) = \mathbb{E}_P(h)$, hence to achieve (8.80), it is sufficient to consider $\widetilde{h}_\varepsilon$ and $\widetilde{h}$ in space $(\widetilde{\Omega}, \widetilde{\mathcal{F}}, \widetilde{P})$ with Q-Wiener process, $\widetilde{W}_t$ in place of h_ε, h and W_t. As in [9], we keep the same notation as (h_ε, h, W_t) for better presentation of the proof.

We have,

$$U^\varepsilon(t) = -\int_0^t AU^\varepsilon(s)ds - \int_0^t \left(B(u_{h_\varepsilon}(s)) - B(u_h(s))\right) ds$$

$$- \int_0^t \left(\widetilde{R}(s, u_{h_\varepsilon}(s)) - \widetilde{R}(s, u_h(s))\right) ds$$

$$+ \int_0^t \left(\sigma(s, u_{h_\varepsilon}(s))h_\varepsilon(s) - \sigma(s, u_h(s))h(s)\right) ds + \sqrt{\varepsilon} \int_0^t \sigma(s, u_{h_\varepsilon}(s))dW(s),$$

which by Itô's formula becomes,

$$
\begin{aligned}
|U^{\varepsilon}(t)|^2 + 2 \int_0^t \|U^{\varepsilon}(s)\|^2 ds = &-2 \int_0^t \big(B(u_{h_\varepsilon}(s)) - B(u_h(s)), U^{\varepsilon}(s) \big) ds \\
&- 2 \int_0^t \big(\tilde{R}(s, u_{h_\varepsilon}(s)) - \tilde{R}(s, u_h(s)), U^{\varepsilon}(s) \big) ds \\
&+ 2\sqrt{\varepsilon} \int_0^t \big(\sigma(s, u_{h_\varepsilon}(s)) dW(s), U^{\varepsilon}(s) \big) \\
&+ 2 \int_0^t \big(\sigma(s, u_{h_\varepsilon}(s)) h_\varepsilon(s) - \sigma(s, u_h(s)) h(s), U^{\varepsilon}(s) \big) ds \\
&+ \varepsilon \int_0^t \|\sigma(s, u_{h_\varepsilon}(s))\|_{L_Q}^2 ds.
\end{aligned}
\tag{8.81}
$$

Under the assumption that the left hand side of equality (8.81) is indeed bounded, the expectation of both sides may be taken and similar to estimates for (8.77), each term on the right may be bounded above to obtain by Gronwall's inequality,

$$
\mathbb{E}\|U^{\varepsilon}(t)\|_{\mathcal{C}([0,T];H) \cap L^2(0,T;V)} \to 0 \quad \text{as} \quad \varepsilon \to 0.
\tag{8.82}
$$

We now verify that the left hand side of (8.81) is indeed bounded. Let,

$$
\begin{aligned}
G_{N,\varepsilon}(t) = &\left\{ \sup_{0 \le s \le t} |u_{h_\varepsilon}(s)|^2 \le N \right\} \cap \left\{ \sup_{0 \le s \le t} |u_h(s)|^2 \le N \right\} \\
&\cap \left\{ \int_0^t \|u_{h_\varepsilon}(s)\|^2 ds \le N \right\} \cap \left\{ \int_0^t \|u_h(s)\|^2 ds \le N \right\}.
\end{aligned}
$$

Recall the Markov and Chebyshev inequalities,

$$
P(|X| > k) \le \frac{\mathbb{E}(|X|)}{k}, \quad \text{and} \quad P(|X| > k) \le \frac{\mathbb{E}(|X|^2)}{k^2},
$$

respectively, for any positive constant, k. Thus, by Markov inequality and (8.72), we estimate the complement of $G_{N,\varepsilon}(T)$ as,

$$
\begin{aligned}
P(G_{N,\varepsilon}(T)^c) \le &\frac{1}{N} \mathbb{E}\left(\sup_{0 \le s \le t} |u_{h_\varepsilon}(s)|^2 \right) + \frac{1}{N} \mathbb{E}\left(\sup_{0 \le s \le t} |u_h(s)|^2 \right) \\
&+ \frac{1}{N} \mathbb{E}\left(\int_0^t \|u_{h_\varepsilon}(s)\|^2 ds \right) + \frac{1}{N} \mathbb{E}\left(\int_0^t \|u_h(s)\|^2 ds \right) \\
\le &\frac{K}{N}.
\end{aligned}
$$

Since the above is true for any $N > 0$, then setting $N \to \infty$, gives, $P(G_{N,\varepsilon}(T)^c)$ $= 0$. Observe that this technique is analogous to the use of stopping time, τ_N in the proof of Theorem 8.1 discussed in Section 8.1. Therefore, Chebyshev inequality yields,

$$P\left(\|U^\varepsilon\|_{\mathcal{C}([0,T];H) \cap L^2(0,T;V)} > \delta\right) \leq \frac{1}{\delta^2}\, \mathbb{E}\left(\|U^\varepsilon\|^2_{\mathcal{C}([0,T];H) \cap L^2(0,T;V)}\right),$$

for any fixed $\delta > 0$ and we obtain (8.80) by (8.82).   □

In [1], H. Bessaih and A. Millet also study the large deviations for two-dimensional incompressible stochastic Navier-Stokes equations with the main difference that they multiply the noise term by v representing viscosity and let $v \to 0$ instead of $\varepsilon \to 0$. Namely, they consider,

$$du^v(t) + \left(vAu^v(t) + B(u^v(t), u^v(t))\right) dt = \sqrt{v}\, \sigma_v(t, u^v(t))dW(t), \qquad (8.83)$$

with S_M defined as in (8.69). Here the controlled PDE and stochastic controlled equations are

$$du_h^0(t) + B(u_h^0(t), u_h^0(t))dt = \tilde{\sigma}_0(t, u_h^0(t))h(t)dt, \qquad (8.84)$$

and

$$du_h^v(t) + \left(vAu_h^v(t) + B(u_h^v(t), u_h^v(t))\right) dt$$
$$= \sqrt{v}\sigma_v(t, u_h^v(t))dW(t) + \tilde{\sigma}_v(t, u_h^v(t))h(t)dt, \qquad (8.85)$$

respectively, where $\sigma_v(t, u)$ and $\tilde{\sigma}_v(t, u)$ satisfy the following conditions,

$$\|\sigma_v(t, u)\|^2_{L_Q} \leq K_1 + K_2|u|^2, \quad \text{and} \quad \|\sigma_v(t, u) - \sigma_v(t, v)\|^2_{L_Q} \leq K_1|u - v|^2,$$

$$\|\tilde{\sigma}_v(t, u)\|_{L(H_0,H)} \leq K_1 + K_2|u|^2, \quad \text{and} \quad \|\tilde{\sigma}_v(t, u) - \tilde{\sigma}_v(t, v)\|_{L(H_0,H)} \leq K|u - v|^2.$$

where $|.|$ denotes the norm in H. Similar to the proofs in [9,15], the authors implement the Galerkin approximation scheme to prove the well-posedness of weak in PDE and strong in probability solutions of the stochastic controlled and controlled PDE equations in the paper's Theorems 2.3 and 3.1, respectively. In this paper, the space of solutions is considered to be $\mathcal{C}(0, T; H) \cap L^\infty(0, T; H^{1,q} \cap V) \cap L^2(0, T; \mathcal{H})$ for $q \in [2, \infty)$, where here $\mathcal{H} := Dom(A^1/4)$. Similar to the techniques presented in this section, the authors verify the first and second conditions required by the weak convergence approach in the article's Propositions 5.4 and 5.3, respectively and obtain the following theorem.

Theorem 8.12 *(Theorem 5.2 in [1]) The family, $\{u^{\nu}\}_{\nu>0}$ of solutions to (8.83) satisfies the large deviations principle in $\mathcal{C}(0, T; H) \cap L^{\infty}(0, T; H^{1,q} \cap V) \cap L^2(0, T; \mathcal{H})$ for $q \in [2, \infty)$ with rate function,*

$$I(u) = \inf_{\{h \in L^2(0,T;H_0) \, : \, u = \mathcal{G}^0(\int_0^{\cdot} h(s)ds)\}} \frac{1}{2} \int_0^T |h(s)|_0^2 ds. \qquad (8.86)$$

8.4 Appendix

8.4.1 Brownian Motion, Lévy Process and Random Walk

Brownian motion, $\{B_t\}_{t\geq 0}$, also called Wiener process, requires each path, $t \to B(t)$, that is $B(t, \cdot)$ independent of ω, to be continuous with respect to t, and the increments, $B(t, \omega) - B(s, \omega)$ for $t > s$ to have normal distribution, $N(0, t - s)$ and be independent. Namely, for $t_0 < t_1 < \ldots < t_n$, we have $B(t_j) - B(t_i) \sim N(0, t_j - t_i)$ for $j > i$ and

$$B(t_1) - B(t_0), \quad B(t_2) - B(t_1), \quad \ldots, \quad B(t_n) - B(t_{n-1}),$$

are independent random variables, where by convention, the shorthand notation $B(t)$ is used for $B(t, \omega)$. A standard Brownian motion requires the increments to have standard normal distribution, $N(0, 1)$.

A Lévy process, $\{L_t\}_{t\geq 0}$, is any collection of random variables that satisfies, $L_0 = 0$ and has increments that are stationary and independent. A random variable, X_t is stationary if for every $u > 0$, $X(t + u)$ has the same distribution as $X(t)$. Thus, the distribution does not change with time. Note that based on this definition, a Brownian motion is also a Lévy process.

Functions that are right continuous with left limits are referred to as càdlàg functions. It is known that every Lévy process, $\{L_t\}_{t\geq 0}$ has a càdlàg modification (see Theorem 4.3 in [28] for a proof), implying that its paths, $t \to L(t, \omega)$ are càdlàg. Thus, this process is often used to describe processes that have jumps in their graphs. A stochastic process, $\{X_t\}_{t\geq 0}$ is a modification of another stochastic process, $\{Y_t\}_{t\geq 0}$, if for every index t, $X_t = Y_t$ almost surely.

A random walk is a collection of partial sums of a family of independent and identically distributed (i.i.d.) random variables. More precisely, a random walk, $\{S_n\}_{n\geq 0}$ starts with $S_0 = 0$ and for each $n \in \mathbb{N}$,

$$S_n = \sum_{i=1}^{n} X_i,$$

where X_is are i.i.d. random variables. One can think of a random walk as an object moving on a grid, in which a particular value is given based on the direction it takes for each step. A simple random walk is in the one-dimension setting and each X_i can only take values 1 or -1. That is, if for the ith step the object moves to the right on the grid, the value will be 1 and if it moves to the left, the value will be -1. Then $|S_n|$ gives the distance the object has moved from 0 after n steps and S_n gives its exact position on the x-axis. Random walk can also be interpreted as the discrete counterpart of the Brownian motion, where the step size is made very small and it can be shown, as it is well-known in the literature, that each random walk converges to a Brownian motion as n is set to go to infinity.

8.4.2 Applications of large deviations

Here we point out some results in finance and queueing theory that apply large deviations to obtain an upperbound to the probability of a parameter in the problem exceeding a given threshold. For other examples of applications we refer the reader to [24,34]. We note that to avoid a distraction from the main topic, we only state the main asymptotic limits determined by large deviations. Most of the papers discussed here use classical theorems in large deviations given in [14].

As for applications in finance, H. Pham in [29] lets X_t represent the wealth of the investor at time t and grouping assets as risky, denoted by S_t or nonrisky, denoted by S_t^0, the author uses the following model of $(X_t)_t$ given by an Itô process,

$$dX_t = X_t \left(\pi_t \frac{dS_t}{S_t} + (1 - \pi_t) \frac{dS_t^0}{S_t^0} \right),$$

where, π_t is the money the investor has put in the risky asset, S_t. The goal is to set a benchmark, denoted by I_t and investigate the ratio of X_t with this benchmark. That is, the author considers $Z_t^\pi = \frac{X_t^\pi}{I_t}$ and letting $L_T^\pi := \frac{\ln Z_T^\pi}{T}$, proves by large deviations,

$$\lim_{n \to \infty} \limsup_{T \to \infty} \frac{1}{T} \log P \left(L_T^{\pi^{*,n}} \geq c \right) = - \sup_{\lambda \in [0, \bar{\lambda}]} (\lambda c - \Lambda(\lambda)),$$

for a constant $c \in \mathbb{R}$ and a sequence $\pi_t^{*,n}$. As for examples of other results in finance, denoting L_n as the total loss and $S_n^{(\alpha)}$ as the profit from n risky assets, limits,

$$\lim_{n \to \infty} \frac{1}{n} \log P \left(L_n > x_n \right) = -\frac{1}{2} \|\gamma_*\|^2,$$

and

$$\limsup_{n\to\infty} \frac{1}{n} \sup_{\alpha\in\mathcal{A}} \log P\left(S_n^{(\alpha)} \geq r\right) = -\Lambda^*(r),$$

were proved in [18] and [8], respectively, using the theory of large deviations.

In the study of queues, in [2] J. Blanchet, X. Chen and H. Lam let $\bar{Q}_\lambda(t,y)$ denote the number of customers who are waiting to be served at time t. The customers are assumed to have an arrival rate of $\lambda > 0$ and y denotes the remaining service time available for all. The authors prove the large deviations principle for $\left\{\frac{\bar{Q}_\lambda}{\lambda}\right\}_{\lambda>0}$. Namely, they prove that for any open set U and closed set C,

$$- \inf_{q\in U} I(q) \leq \liminf_{\lambda\to\infty} \frac{1}{\lambda} P\left(\frac{\bar{Q}_\lambda}{\lambda} \in U\right) \leq \limsup_{\lambda\to\infty} \frac{1}{\lambda} P\left(\frac{\bar{Q}_\lambda}{\lambda} \in C\right) \leq - \inf_{q\in C} I(q).$$

Considering another type of queue, in which customers are assumed to be served at random, B. Kim and J. Kim in [21], let G represent a busy period and W_{ROS} be the waiting time of any arbitrary customer and prove the following limits,

$$\lim_{t\to\infty} -\frac{1}{t} \log P(G > t) = -\kappa(\theta_*),$$

$$\lim_{t\to\infty} -\frac{1}{t} \log P(W_{ROS} > t) = -\kappa(\theta_*).$$

References

1. H. Bessaih and A. Millet (2009). Large deviation principle and inviscid shell models. *Elec. J. Probabl.* vol. 14, no. 89, 2551–2579.

2. J. Blanchet, X. Chen and H. Lam (2014). Two-parameter sample path large deviations for infinite-server queues. *Stoch. Syst.* vol. 4, no. 1, 206–249.

3. D. Breit, E. Feireisl, M. Hofmanová (2018). Local strong solutions to the stochastic compressible Navier-Stokes system. *Comm. Partial Diff. Eq.* vol. 43, no. 2, 313–345.

4. A. Budhiraja, P. Dupuis (2000). A variational representation for positive functionals of infinite-dimensional Brownian motion. *Probab. Math. Stat.* vol. 20, 39–61.

5. A. Budhiraja, P. Dupuis (2019). *Analysis and Approximation of Rare Events: Representations and Weak Convergence Methods*. Probability Theory and Stochastic Modelling, vol. 94, Springer, New York.

6. A. Budhiraja, P. Dupuis and V. Maroulas (2008). Large deviations for infinite dimensional stochastic dynamical systems. *Ann. Probab.* vol. 36, no. 4, 1390–1420.

7. P. Chow and J. Menaldi (1990). Exponential estimates in exit probability for some diffusion processes in Hilbert spaces. *Stoch. Stoch. Reports.* vol. 29, 377–393.

8. B. Chu (2012). Large deviations estimation of the windfall and shortfall probabilities for optimal diversified portfolios. *Ann. Finance.* vol. 8, 97–122.

9. I. Chueshov and A. Millet (2010). Stochastic 2D Hydrodynamical type systems: well posedness and large deviations. *Appl. Math. Optim.* vol. 61, 379–420.

10. D. Crisan, F. Flandoli, and D. Holm (2019). Solution properties of a 3D stochastic Euler fluid equation. *J. Nonlinear Sci.* vol. 29, 813–870.

11. G. Da Prato and J. Zabczyk (2014). *Stochastic Equations in Infinite Dimensions.* Encyclopedia of Mathematics and Its Applications, vol. 152, Cambridge University Press, Cambridge, United Kingdom.

12. P.A. Davidson (2017). *Introduction to Magnetohydrodynamics: Second Edition.* Cambridge Texts in Applied Mathematics, Cambridge University Press, Cambridge, United Kingdom.

13. L. Debbi (2016). Well-posedness of the multidimensional fractional stochastic Navier-Stokes equations on the torus and on bounded domains. *J. Math. Fluid. Mech.* vol. 18, 25–69.

14. A. Dembo and O. Zeitouni (2010). *Large Deviations Techniques and Applications: Second Edition.* Springer Applications of Mathematics, vol. 38, Springer, New York, NY.

15. J. Duan and A. Millet (2009). Large deviations for the Boussinesq equations under random influences. *Stoch. Proc. Appl.* vol. 119, 2052–2081.

16. P. Dupuis and R. Ellis (1997). *A Weak Convergence Approach to the Theory of Large Deviations.* Wiley Series in Probability and Statistics. John Wiley & Sons, New York, NY.

17. E. Feireisl, T. Karper and M. Pokorný (2016). *Mathematical Theory of Compressible Viscous Fluids: Analysis and Numerics.* Lecture Notes in Mathematical Fluid Mechanics, Birkhäuser, Springer, New York.

18. P. Glasserman, W. Kang and P. Shahabuddin (2007). Large deviations in multifactor portfolio credit risk. *Math. Finance.* vol. 17, no. 3, 345–379.

19. G. Hu and K. Wang (2011). Existence and uniqueness theorem for stochastic differential equations with self-exciting switching. *Discrete Dyn. in Nature Soci.* vol. 2011, 1–12.

20. G. Kallianpur and J. Xiong (1995).*Stochastic Differential Equations in Infinite Dimensional Spaces.* Institute of Mathematical Statistics Lecture Notes- Monograph Series, vol. 26.

21. B. Kim and J. Kim (2013). Large deviations of the waiting time in the GI/G/1 queue with random order service. *Queueing Syst.* vol. 74, 431–443.

22. F. Klebaner (2005). *Introduction to Stochastic Calculus with Applications: Second Edition.* Imperial College Press, London, United Kingdom.

23. H. Kuo (2006). *Introduction to Stochastic Integration.* Springer Universitext, Springer, New York, NY.

24. M. Mandjes (2007). *Large Deviations for Gaussian Queues: Modelling Communication Networks*. John Wiley & Sons, England.

25. D. Nualart and C. Rovira (2000). Large deviations for stochastic Volterra equations. *Bernoulli*. vol. 6, no. 2, 339–355.

26. B. Oksendal (2005). *Stochastic Differential Equations: An Introduction with Applications: Sixth Edition*. Springer Universitext, Springer, Berlin, Germany.

27. A. Pazy (1983). *Semigroups of Linear Operators and Applications to Partial Differential Equations*. Springer Applied Mathematical Sciences, vol. 44, Springer, New York.

28. S. Peszat and J. Zabczyk (2007). *Stochastic Partial Differential Equations with Lévy Noise*. Encyclopedia of Mathematics and its Applications, vol. 113, Cambridge Univeristy Press, Cambridge, United Kingdom.

29. H. Pham (2003). A large deviations approach to optimal long term investment. *Finance Stoch*. vol. 7, 169–195.

30. P. Plotnikov and J. Sokolowski (2012). *Compressible Navier-Stokes Equations: Theory and Shape Optimization*. Monografie Matematyczne, vol. 73, Birkhäuser, Springer, New York.

31. H. Sohr (2001). *The Navier-Stokes Equations: An Elementary Functional Analytic Approach*. Modern Birkhäuser Classics, Verlag, Switzerland.

32. S. Sritharan and P. Sundar (2006). Large deviations for the two-dimensional Navier-Stokes equations with multiplicative noise. *Stoch. Proc. Appl*. vol. 116, 1636–1659.

33. R. Temam (2001). *Navier-Stokes Equations: Theory and Numerical Analysis*. AMS Chelsea Publishing, American Mathematical Society, Providence, Rhode Island.

34. A. Vulpiani, F. Cecconi, M. Cencini, A. Puglisi, and D. Vergni (Editors) (2014). *Large Deviations in Physics: The Legacy of the Law of Large Numbers*. Spinger Lecture Notes in Physics, vol. 885, Springer, New York.

35. R. Wang, J. Zhai, and T. Zhang (2015). A moderate deviation principle for 2-D stochastic Navier-Stokes equations. *J. Differential Equations*. vol. 258, 3363–3390.

36. E. Zeidler (1990). *Nonlinear Functional Analysis and its Applications II/A: Linear Monotone Operators*. Springer, New York, NY.

Index

A

admissible control, 99
Azencott method, 182

B

backscattering, 84
biharmonic Schrödinger equation, 92
Brownian motion, 201
Burkholder-Davis-Gundy inequality, 172

C

central limit theorem, 170
compressible Navier-Stokes equation, 161
correlation function, 118
cost functional, 99
cylindrical Brownian motion, 165

D

Dangling node, 129
defocusing Schrödiner equation, 88
diffraction, 84
discrete Fourier transform, 117
dispersion, 84
dynamic programming, 100

E

eddies, 85

F

focusing Schrödinger equation, 88
forward scattering, 84
Fréchet derivative, 104

G

Gagliardo-Nirenberg inequality, 96

Gâteaux derivative, 104
Girsanov's transformation theorem, 184
Google matrix, 130
Gronwall inequality, 125

H

Hamiltonian equation, 100
homogeneous Sobolev space, 124

I

irradiance, 84
Itô formula, 162–164
Itô process, 162

K

Kerr medium, 88
Kerr nonlinearity, 88
Kolmogorov power spectrum, 118

L

large deviations, 180–181
Laurent Series expansion of Markov chains, 149
Lévy process, 201
Link spamming, 129, 135

M

Markov Chain, 155–158
maximal solution, 167
mild solution, 166
M-Matrix, 144

O

optimal control, 98–100

P

PageRank, 129
pathwise uniqueness, 167
phase screen, 86–87, 117–123
Poincaré inequality, 95
Power method, 130

Q

Q-Wiener process, 165

S

scattering theory, 125
Schwartz space, 123
self-focusing of laser, 83
Sherman-Morison Formula, 140
Skorohod representation theorem, 193
Snell's law, 86
Spam Farms, 154
SpamRank, 154
stationary probability distribution,
    158–159
stochastic Navier-Stokes equation, 162

**Stochastic process, 154
Stratonovich integral, 164
Strichartz estimates, 88

R

Random walk, 201
recurrent state, 157
refractive index, 85
Reynolds number, 85

T

Tatarski spectrum, 119
transient state, 158
transition probability, 157–158

V

Von Kármán spectrum, 119

W

web communities, 153

Printed in the United States
by Baker & Taylor Publisher Services